Leonard

Fractions
Five Pages Each

Addition
Subtraction
Multiplication
Division

5TH – 9TH

AF718001

Designs in Math

Written & Illustrated By
Randy L. Womack, M.Ed.
Learning Disabilities & Art

Published By

G.E.C. Publications golden educational center

"Leading The Way In Creative Educational Materials" ™

P.O. Box 2593 • Redding, California 96099
1-800-800-1791

TEACHER – PARENT INSTRUCTIONS FOR *Designs in Math*

This book has been written to aid the classroom teacher or parent in activities that reinforce perceptual-motor abilities as well as giving students practice solving basic math problems. The best time to introduce *Designs in Math: Fractions* is after children completely understand the activity's particular math function and need practice solving problems. The pages are organized in a suggested developmental sequence. However, the activities may be used in the order that best meets your students' needs.

Please note on each page is an example problem showing the student how to complete the type of problem on the page (i.e., addition of mixed numbers). As the instructor, please take liberties to modify the process to meet your individual students' needs.

The directions on each page have been simplified in a step–by–step format so students can easily read and understand them — independent of teacher instruction. However, it is recommended that teachers spend some initial teaching time explaining the directions.

At this level most students have mastered the ability to draw using a straight edge. However, if you notice any of your students having difficulty it might be to their advantage to begin with a book from the *Creating Line Design* series. (See the back page of this book for examples of the *CLD* series.)

CONTENTS

PUBLISHED BY GOLDEN EDUCATIONAL CENTER
P.O. BOX 2593
REDDING, CALIFORNIA 96099

ISBN 1–56500–008–0

Tips on Using a Straight Edge

In teaching children how to use a ruler or straight edge, it is important that the following points be included:

1. The child should hold his hand out with the palm down on the table.
2. He should hold the straight edge with the tips of his thumb and fingers — as far apart as is comfortable. (Note: Most children will hold it in the middle, and the ends will move as they draw their lines.)
3. The child should hold his pencil at a perpendicular angle. (Note: Many children will place the pencil point at a 45–30° angle into the edge of the ruler and usually lift the ruler up with the point of the pencil.)
4. Be sure the child places his pencil on one of the dots before he moves the straight edge to that dot. After he has his pencil on one of the dots, he can slide the other end of the straight edge to the dot he wants to connect. This is an essential step to be learned by the student. It is the most accurate way to line up the straight edge with the two dots needing to be connected — and accuracy is an important element in the child's feeling successful about his design.
5. Always have the child move his pencil from top to bottom and from the left to the right. (Left handed children would move from top to bottom and then right to left.) Instruct him to pick his pencil up and place it on the top or left-side dot after the ruler is in place. This will add to the accuracy of the lines. His lines will also be easier to make and neater in appearance.
6. Instruct the child to press gently on the pencil. (Note: Most children will want to press down hard, which causes the straight edge to move, as well as causing hand fatigue.)
7. Some children may reach over the straight edge to connect the dots, using the awkward edge. Instruct them to draw on the right and top edge of the ruler if they are right handed. (A left-handed child should only draw on the left and top edge of the ruler.)
8. To avoid reaching over the ruler, the student may turn his paper, instead of the straight edge, to a comfortable orientation to facilitate his drawing the designs.
9. Some children (usually younger) may find it easier to stand up while working on their designs.
10. The child must always complete connecting the dots in numerical order then alphabetical order. If he does not, the designs will not work and he will get frustrated to no end — but that is part of the learning process.

✎ Please Note: We wrote the above information referring to a male student. We like girls, too. However, we thought it easier reading to use one gender rather than "he/she".

Some Students May Want to Do One or More of the Following Activities

Suggestions for Additional Activities

1. Children enjoy using felt pens for coloring the designs. However, every so often, challenge the students to experience different media such as colored pencils, chalk, "artista" oils and even water color paints.
2. Most children will want to use many colors or just certain favorites. Encourage each student to experiment with color patterns of only two or three colors— a great deal of color wheel reinforcement is possible.
3. After he colors the completed design, have the child re-do the lines with a black pen (when the medium permits).
4. Have the child cut completed designs out and paste them on colored background paper. He could even frame the finished product(s).
5. After he colors a design, have the child cut out the specific shapes and paste them onto a colored piece of paper in the exact design seen on the respective "correction key." Many of the designs are difficult. Be certain that the child is capable of doing the chosen design so he doesn't get frustrated.
6. After he colors a design, have the child cut out the specific shapes and paste them onto a colored piece of paper to create a unique design of his own.
7. **Introduce students to "string art":**
   a. Several of the designs in this book lend themselves to the craft of "string art."
   b. Direct the student to choose one of his favorite designs. (Use your judgement as to the difficulty factor involved.) Have the child complete the design on paper. He does not need to color it.
   c. Give the child a square piece of three-quarter inch thick soft wood (at least 8" square). Plywood is inexpensive and easy to use. It also looks nice when painted or stained a color.
   d. Tape his completed design to the wood with masking tape.
   e. Choose the appropriate nails to work with. (1"–1½" long work well)
   f. Now, direct the student to hammer nails so that about half of each nail remains above the board. Make certain that the nails are placed directly over the dots on the paper taped to the wood.
   g. Have the student tie and run string from one nail to the other until the design is completed. (The stringing may have to be altered from the exact design in order to work. However, encourage the child to allow for design alterations in order to make the stringing easier.)
   h. After the stringing is completed, carefully remove the paper taped to the wood and you have a fine example of "string art."

8. **Introduce students to "nail sculpture":**

   a. Several of the designs in this book lend themselves to the craft of "nail sculpture."

   b. Direct the student to choose one of his favorite designs. (Use your judgement as to the difficulty factor involved.) Have the child complete the design on paper. He does not need to color it.

   c. Give the child a square piece of soft wood (at least 8" square). It must be at least three-fourths inch thick. However, one and one-half inch thick works very well. Try using a rough-sawn exterior wood such as cedar. (Plywood does not work well for this project unless it is hung indoors.) Staining or painting the wood before it is used is optional.

   d. Tape his completed design to the wood with masking tape.

   e. Choose a variety of different nails and screws to use. Make sure to choose some that will rust and some that will not rust (galvanized do not rust.) Get a variety of lengths and head sizes. Brass screws are expensive, but they add a special dimension to the sculpture when it is completed.

   f. Now, direct the student to hammer the nails (or screws), filling an entire shape within the design. (Screws can be hammered very easily. However, screwing them in is excellent for muscle development. He may need instructions on how to make a starter hole with a nail or "Yankee" screwdriver.) Explain that it is important to put the nails as close together as possible in order to create the design effectively. (Some of the shapes within the design may be omitted or changed as needed.)

   g. It is important for the child to use the same type of nail for each of the specific shapes chosen. However, it is equally important for him to vary the different nails within the whole design. The student should also be instructed to vary the depth he hammers the nails into the wood from shape-to-shape. Graduated depths within the same shape creates a very nice result.

   h. After the child has filled the desired shapes (some may be left with just the wood showing), get as much of the paper out as possible and have the child hang his finished sculpture outside to be rusted and changed by the weather. The paper will eventually deteriorate.

9. Some students may want to create their designs and have other students complete them.

CORRECTION KEY

Designs in Math - Add Fractions ...Page 1

***Use this design as an answer key.**

***Allow your students to correct their own work.**
Make a transparency of this design and instruct your students to place the transparency over their completed design for a quick and easy check.

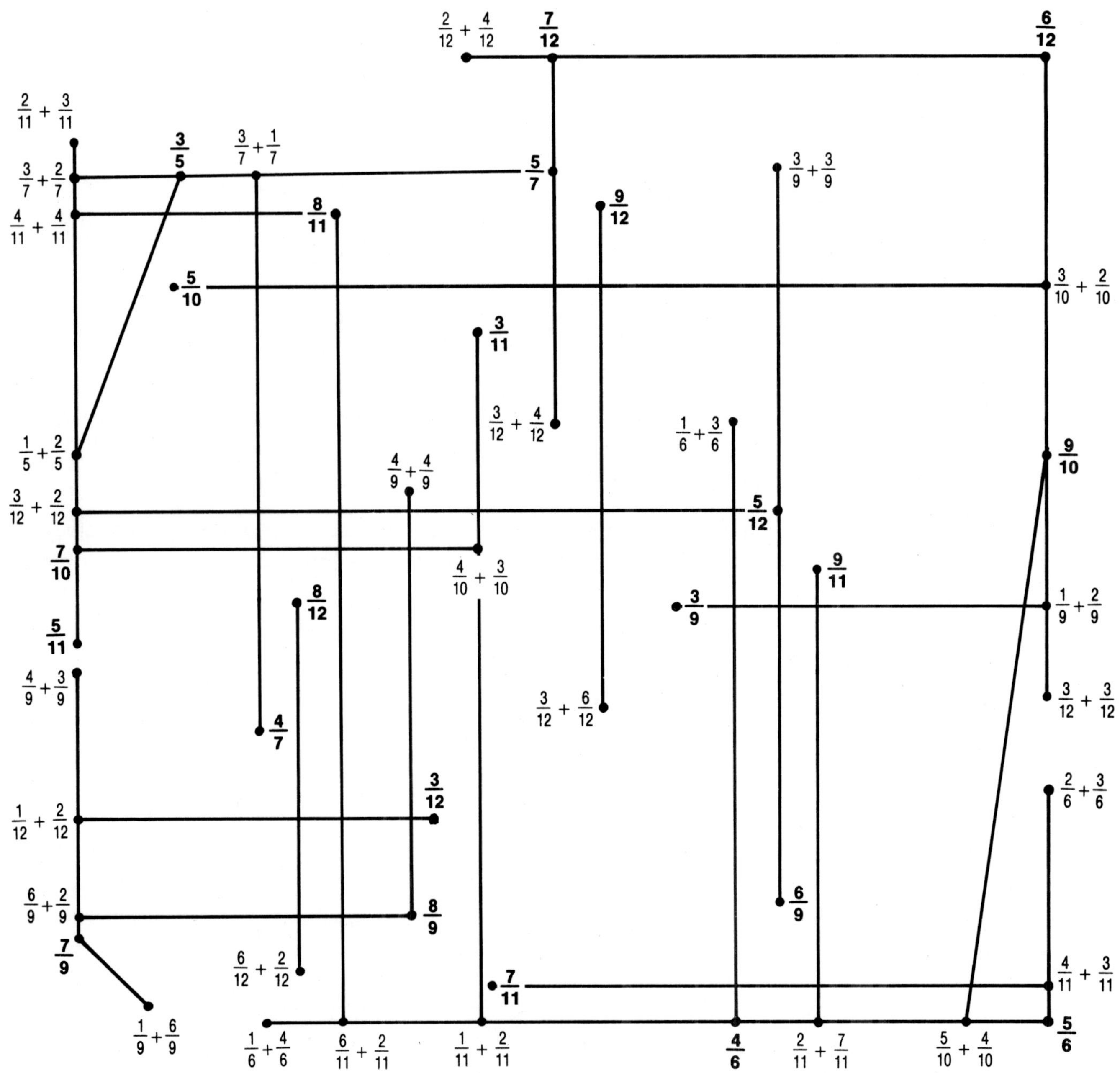

Name ______________________ Date ____________

Read all of the directions before you begin.

1. **Choose one** of the **problems** next to a dot.
2. **Determine** the **sum** of the two fractions.
3. **Use a straight edge** and draw a line to the **correct answer**. Some answers may be used more than once.
4. **Make sure** you do **all** of the problems. Some lines pass through dots that are **not** part of the one being completed.
5. **You may** want to **color** or **shade your design** after you have completed the problems.

EXAMPLE: *Addition of fractional numbers with like denominators.* $\frac{2}{6} + \frac{3}{6} = \frac{2+3}{6} = \frac{5}{6}$

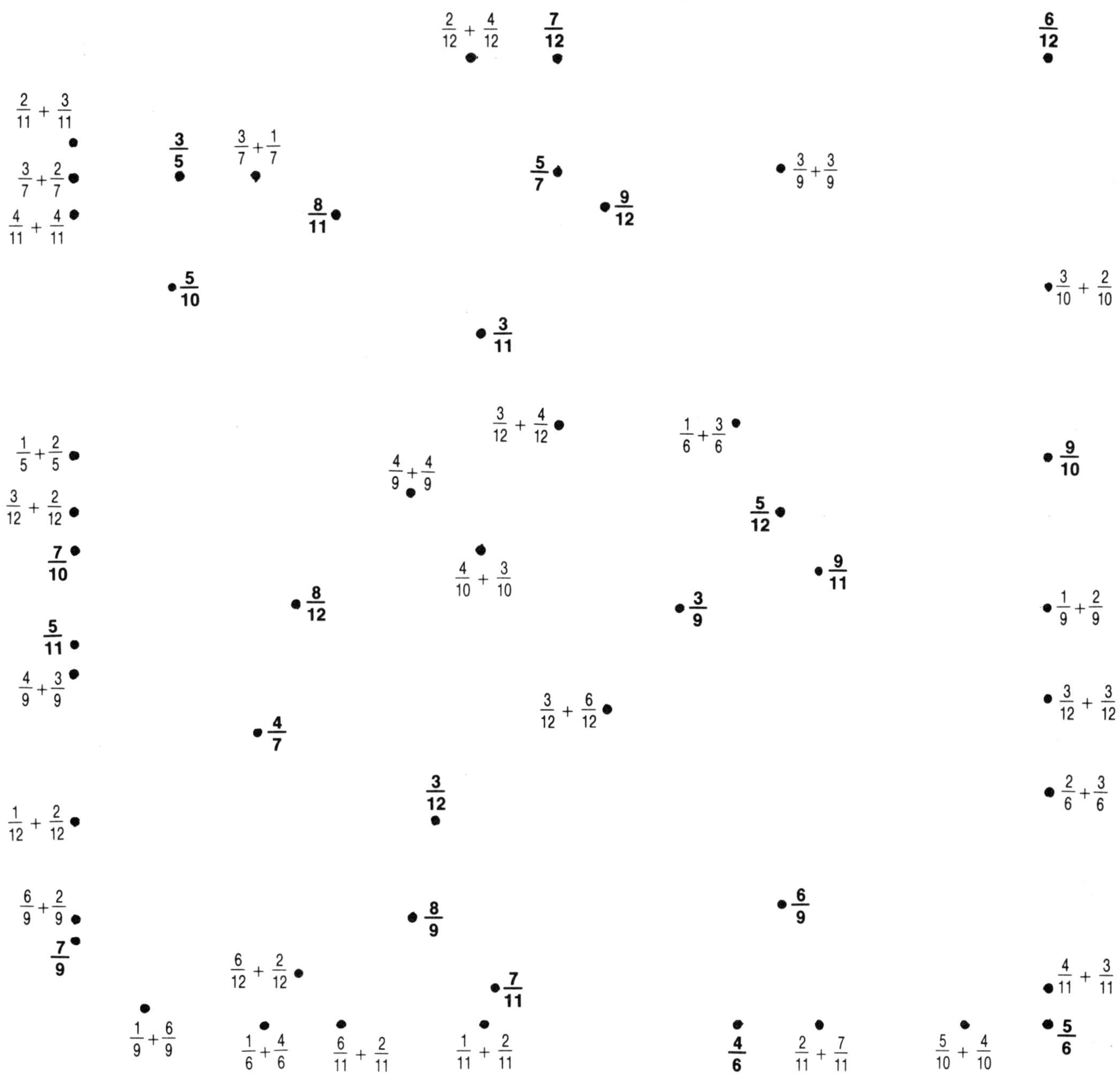

CORRECTION KEY

Designs in Math - Add Fractions ...Page 2

***Use this design as an answer key.**

***Allow your students to correct their own work.**
Make a transparency of this design and instruct your students to place the transparency over their completed design for a quick and easy check.

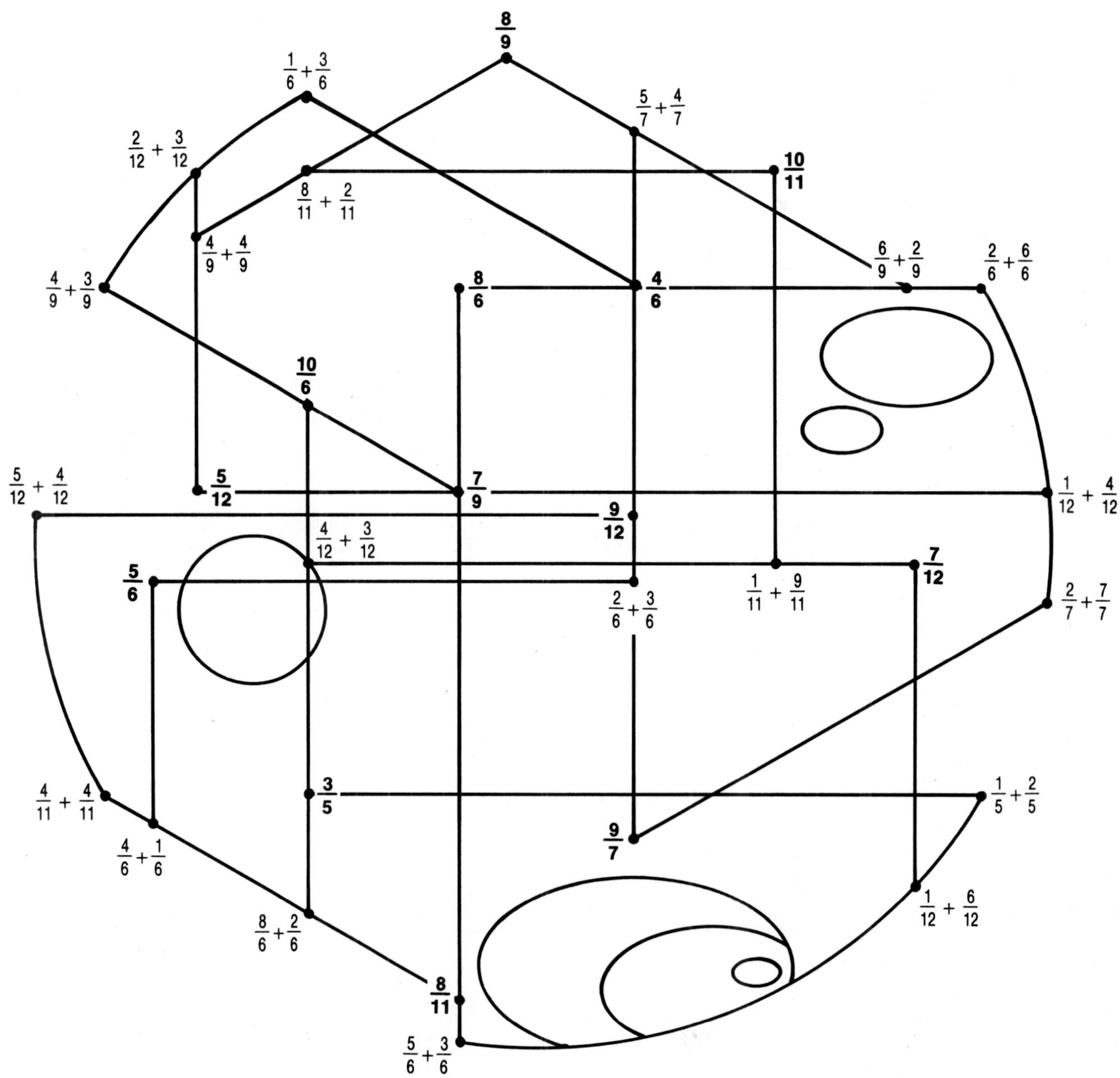

Name ______________________ **Date** ______________

Read all of the directions before you begin.

1. **Choose one** of the **problems** next to a dot.
2. **Determine** the **sum** of the two fractions.
3. **Use a straight edge** and draw a line to the **correct answer**. Some answers may be used more than once.
4. **Make sure** you do **all** of the problems. Some lines pass through dots that are **not** part of the one being completed.
5. **You may** want to **color** or **shade your design** after you have completed the problems.

EXAMPLE: *Addition of fractional numbers with like denominators.* $\frac{2}{6} + \frac{3}{6} = \frac{2+3}{6} = \frac{5}{6}$

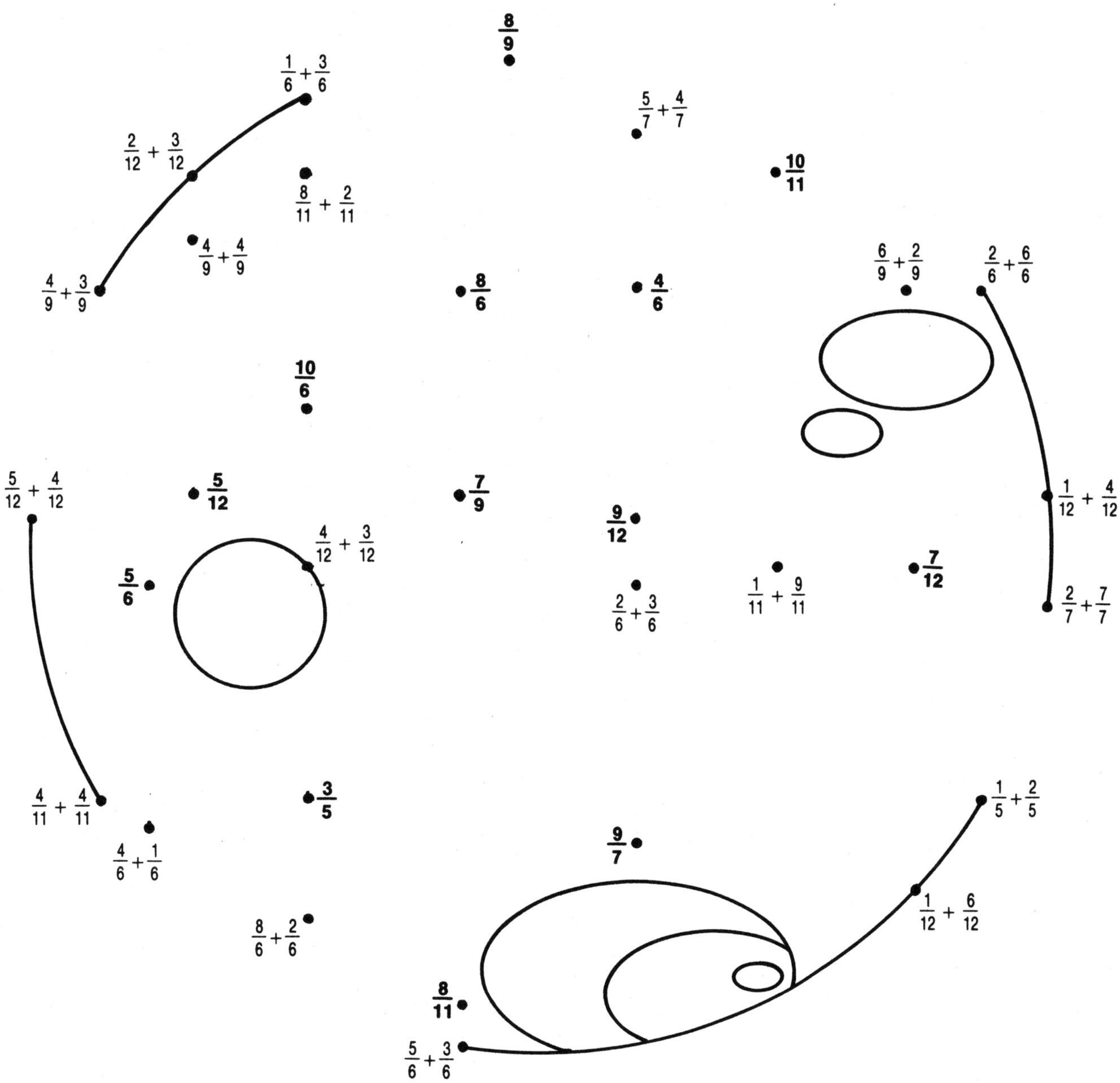

CORRECTION KEY

Designs in Math - Add Fractions ... Page 3

***Use this design as an answer key.**
***Allow your students to correct their own work.**
Make a transparency of this design and instruct your students to place the transparency over their completed design for a quick and easy check.

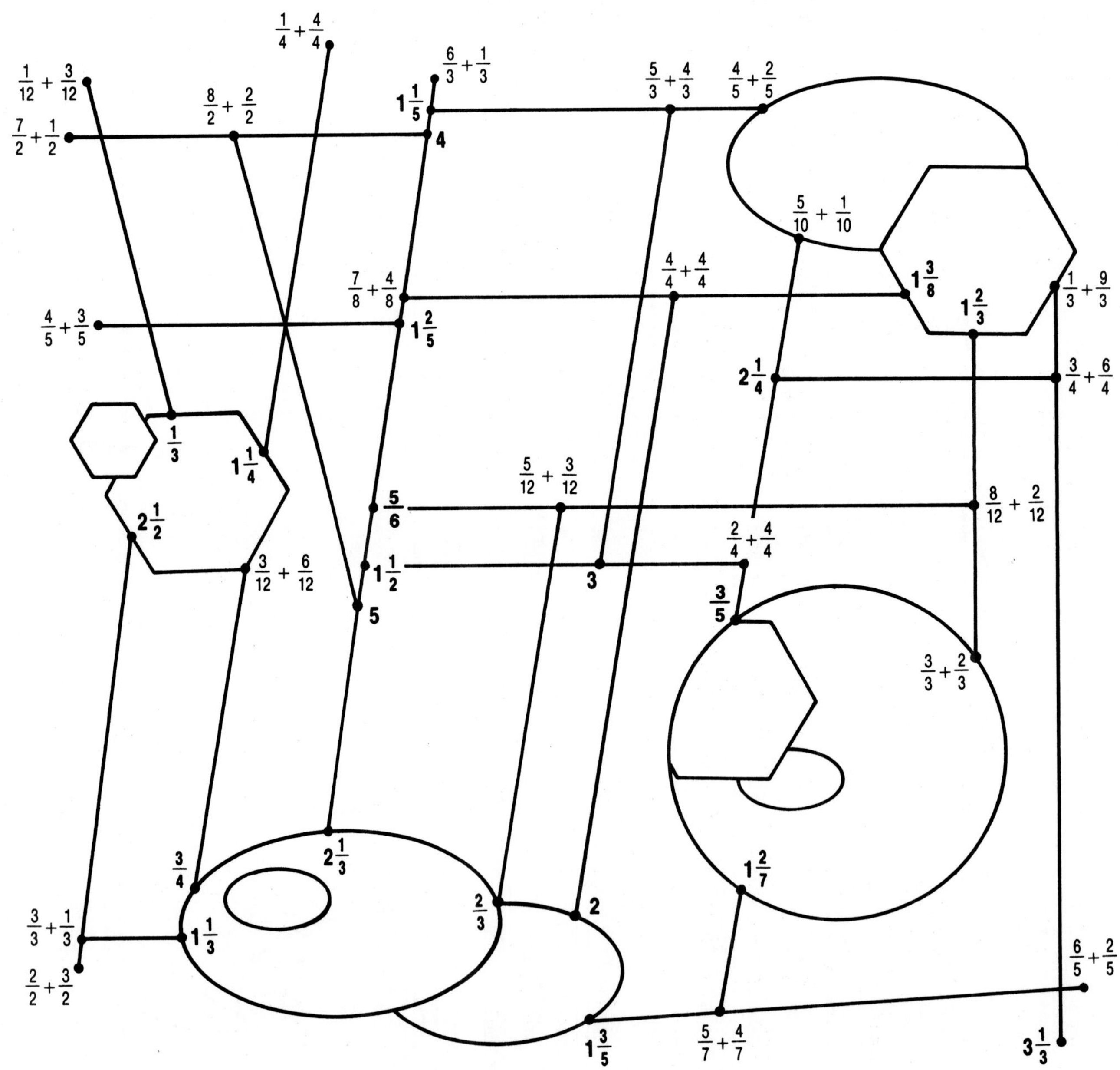

Name ______________________________ Date ______________

Read all of the directions before you begin.

1. **Choose one** of the **problems** next to a dot.
2. **Determine** the **sum** of the two fractions.
3. **Use a straight edge** and draw a line to the **correct answer**.
   Some answers may be used more than once.
4. **Make sure** you do **all** of the problems. Some lines pass through dots that are **not** part of the one being completed.
5. **You may** want to **color** or **shade your design** after you have completed the problems.

***NOTE:** The answers are **reduced** to their **simplest form**.

EXAMPLE: *Addition of fractional numbers with like denominators.* $\frac{2}{6} + \frac{2}{6} = \frac{2+2}{6} = \frac{4}{6} = \frac{2}{3}$

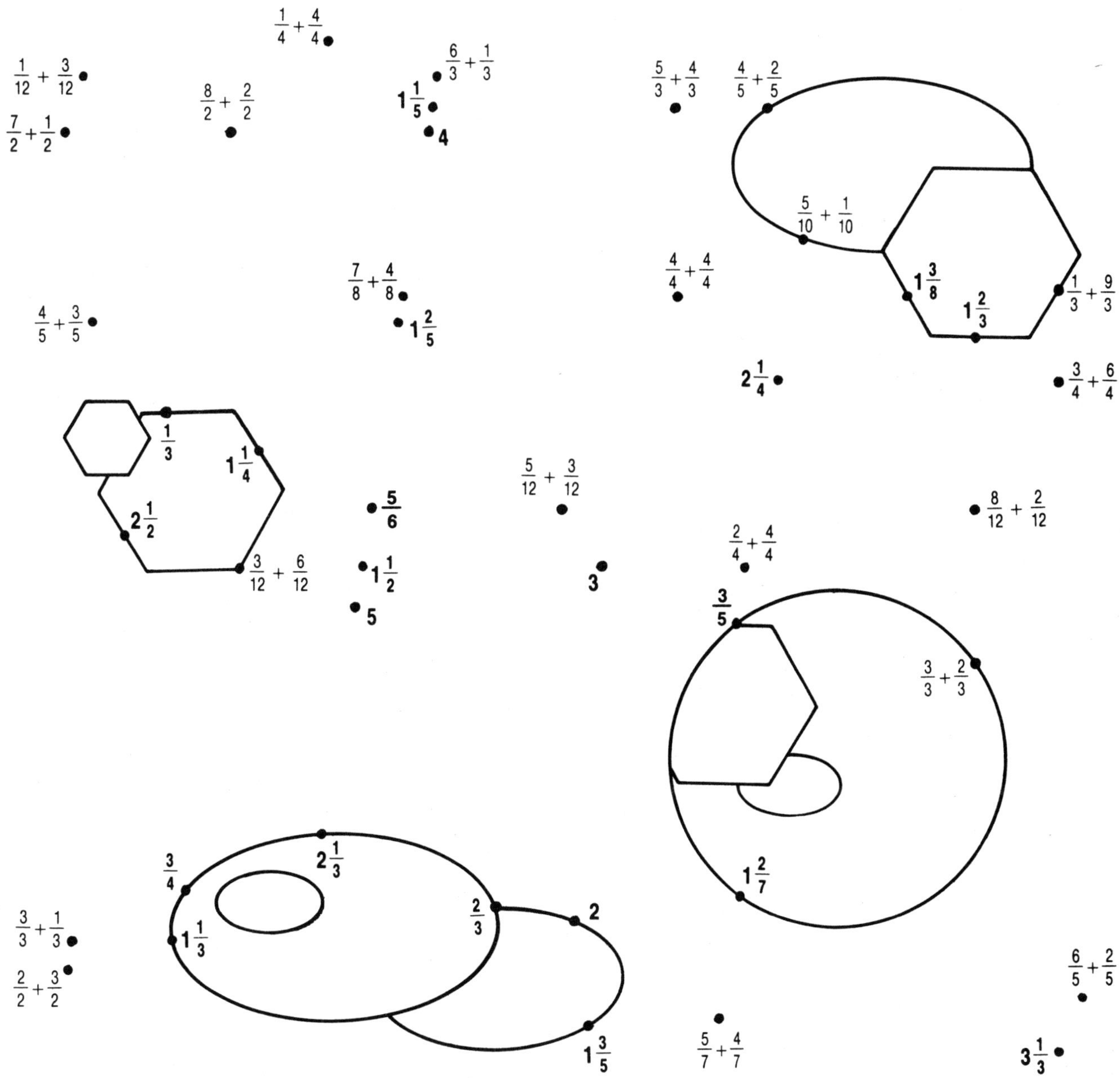

CORRECTION KEY

Designs in Math - Add Fractions ... Page 4

***Use this design as an answer key.**
***Allow your students to correct their own work.**
Make a transparency of this design and instruct your students to place the transparency over their completed design for a quick and easy check.

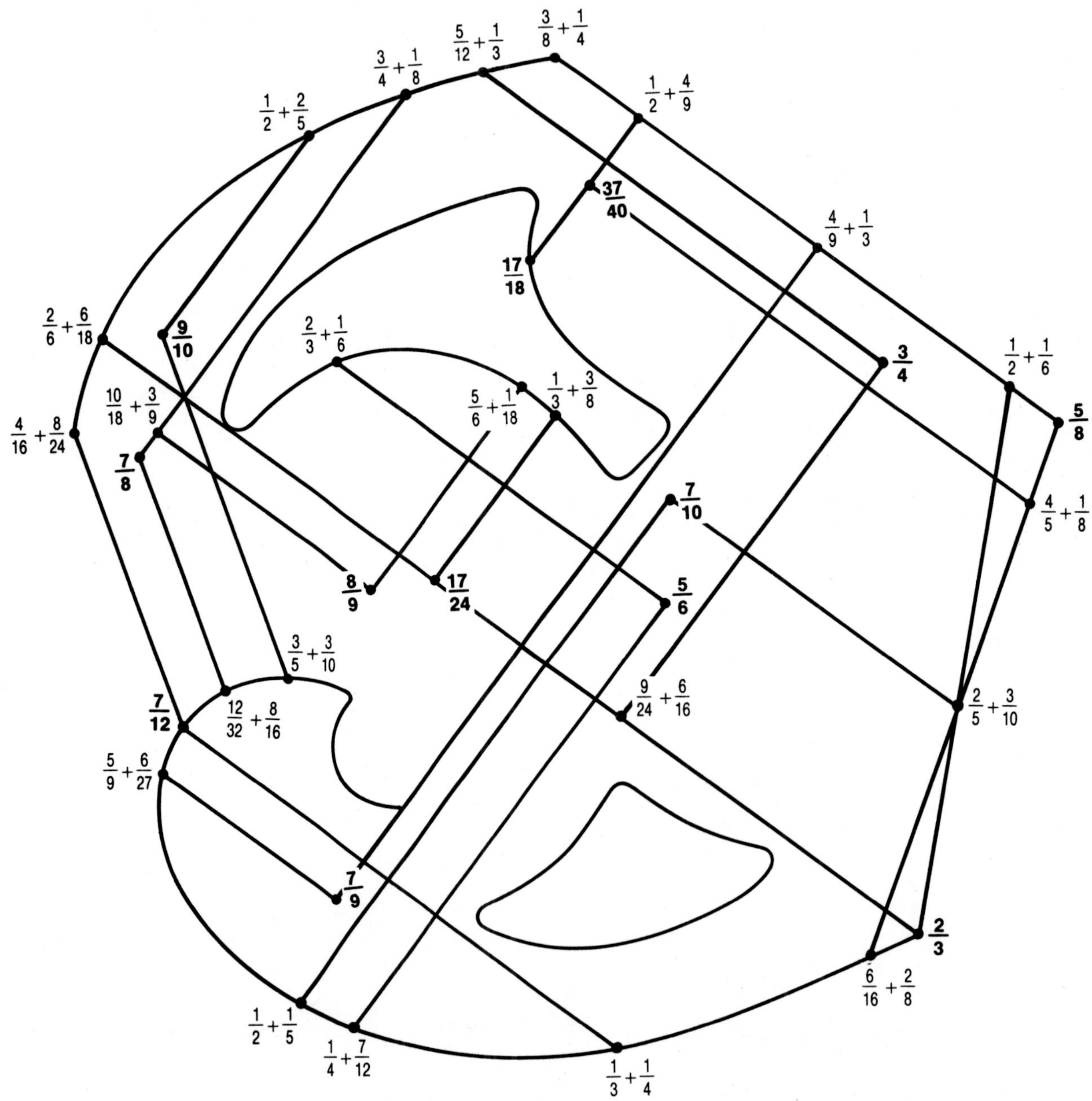

Name ______________________________ Date ______________

Read all of the directions before you begin.

1. **Choose one** of the **problems** next to a dot.
2. **Determine** the **sum** of the two fractions.
3. **Use a straight edge** and draw a line to the **correct answer**. Some answers may be used more than once.
4. **Make sure** you do **all** of the problems. Some lines pass through dots that are **not** part of the one being completed.
5. **You may** want to **color** or **shade your design** after you have completed the problems.

***NOTE:** The answers are **reduced** to their **simplest form**.

***EXAMPLE:** Addition of fractional numbers with unlike denominators.*

$$\frac{4}{8} + \frac{1}{4} =$$

$$\frac{4}{8} + \frac{2}{8} = \frac{4+2}{8} = \frac{6}{8} = \frac{3}{4}$$

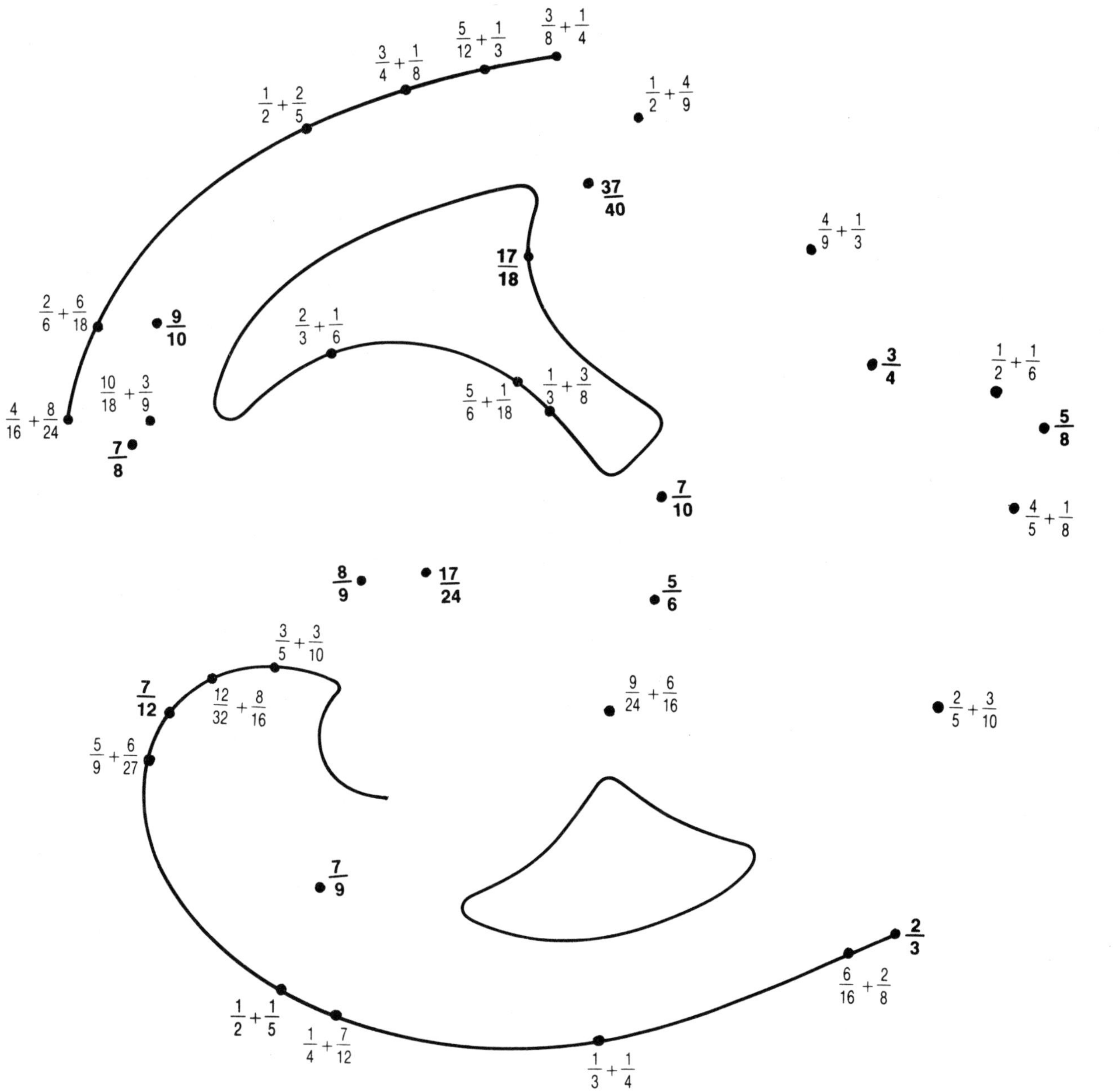

CORRECTION KEY

Designs in Math - Add Fractions ... Page 5

***Use this design as an answer key.**
***Allow your students to correct their own work.**
Make a transparency of this design and instruct your students to place the transparency over their completed design for a quick and easy check.

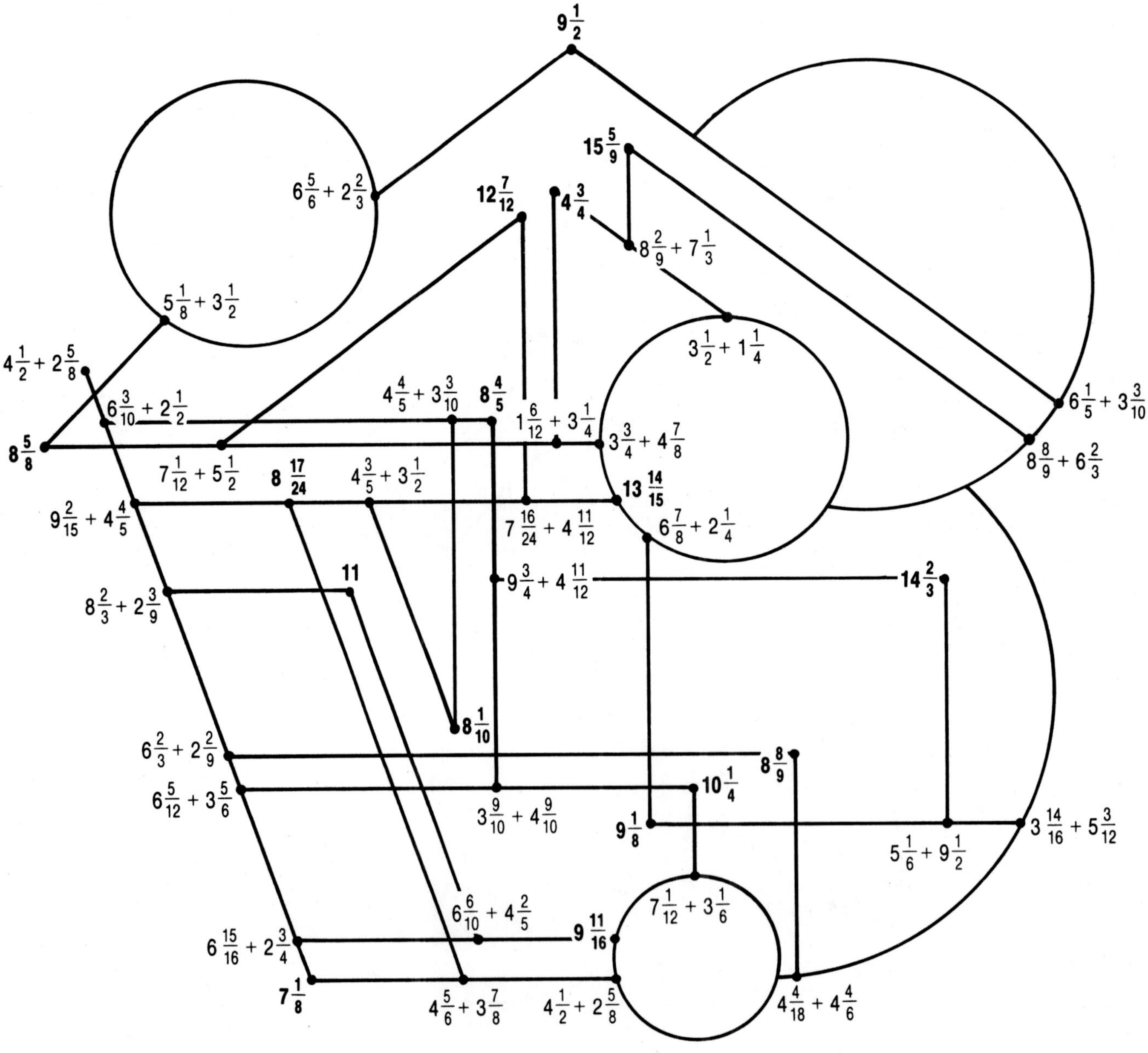

Name ______________________________ **Date** ______________

Read all of the directions before you begin.

1. **Choose one** of the **problems** next to a dot.
2. **Determine** the **sum** of the two fractions.
3. **Use a straight edge** and draw a line to the **correct answer**.
   Some answers may be used more than once.
4. **Make sure** you do **all** of the problems. Some lines pass through dots that are **not** part of the one being completed.
5. **You may** want to **color** or **shade your design** after you have completed the problems.

***NOTE:** The answers are **reduced** to their **simplest form**.

EXAMPLE: *Addition of mixed numbers.* $1\frac{4}{8} + 2\frac{3}{4} = 1\frac{4}{8} + 2\frac{6}{8} = 3\frac{10}{8} = 4\frac{2}{8} = 4\frac{1}{4}$

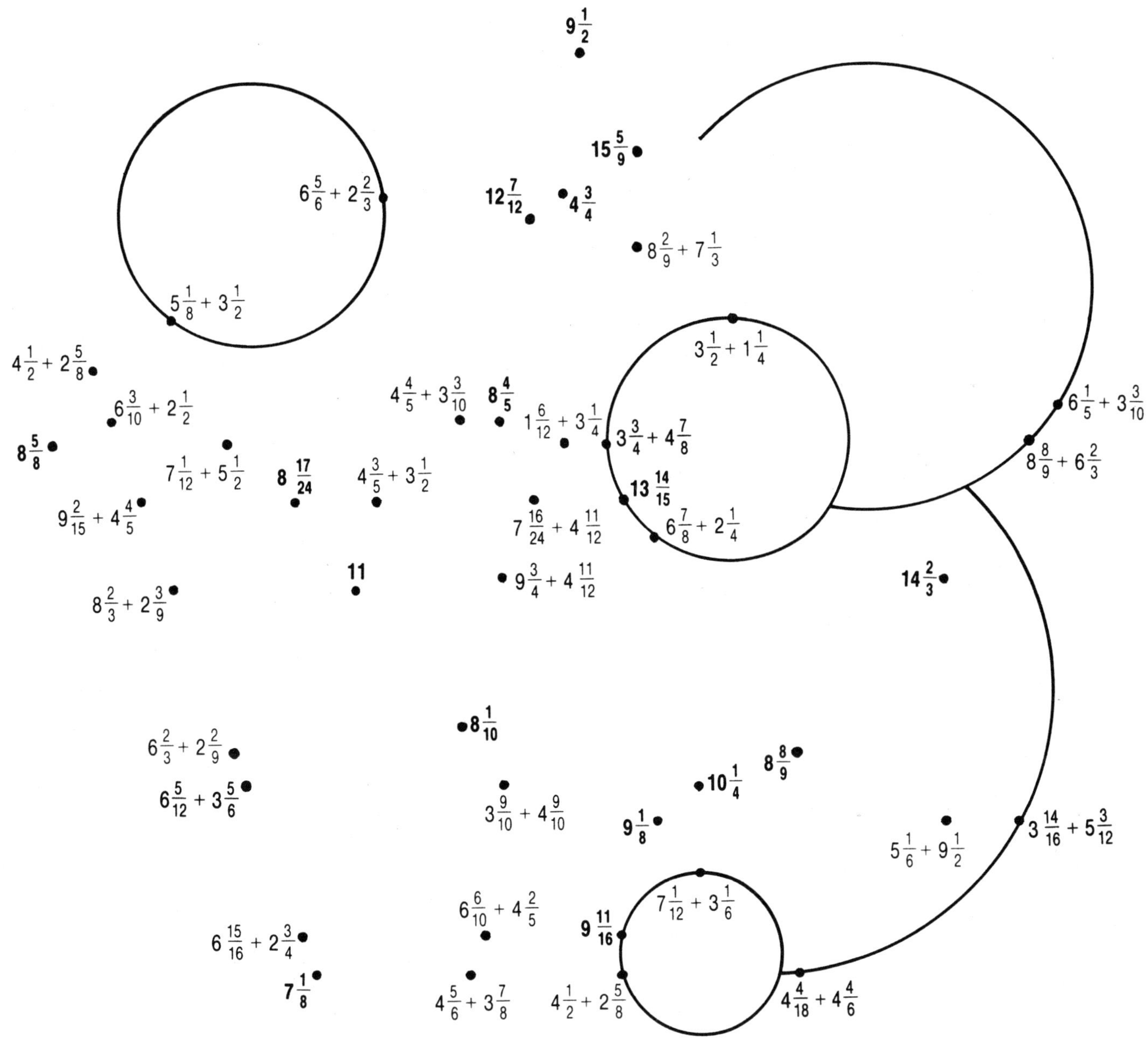

CORRECTION KEY

Designs in Math - Subtract Fractions ... Page 6

***Use this design as an answer key.**

***Allow your students to correct their own work.**
Make a transparency of this design and instruct your students to place the transparency over their completed design for a quick and easy check.

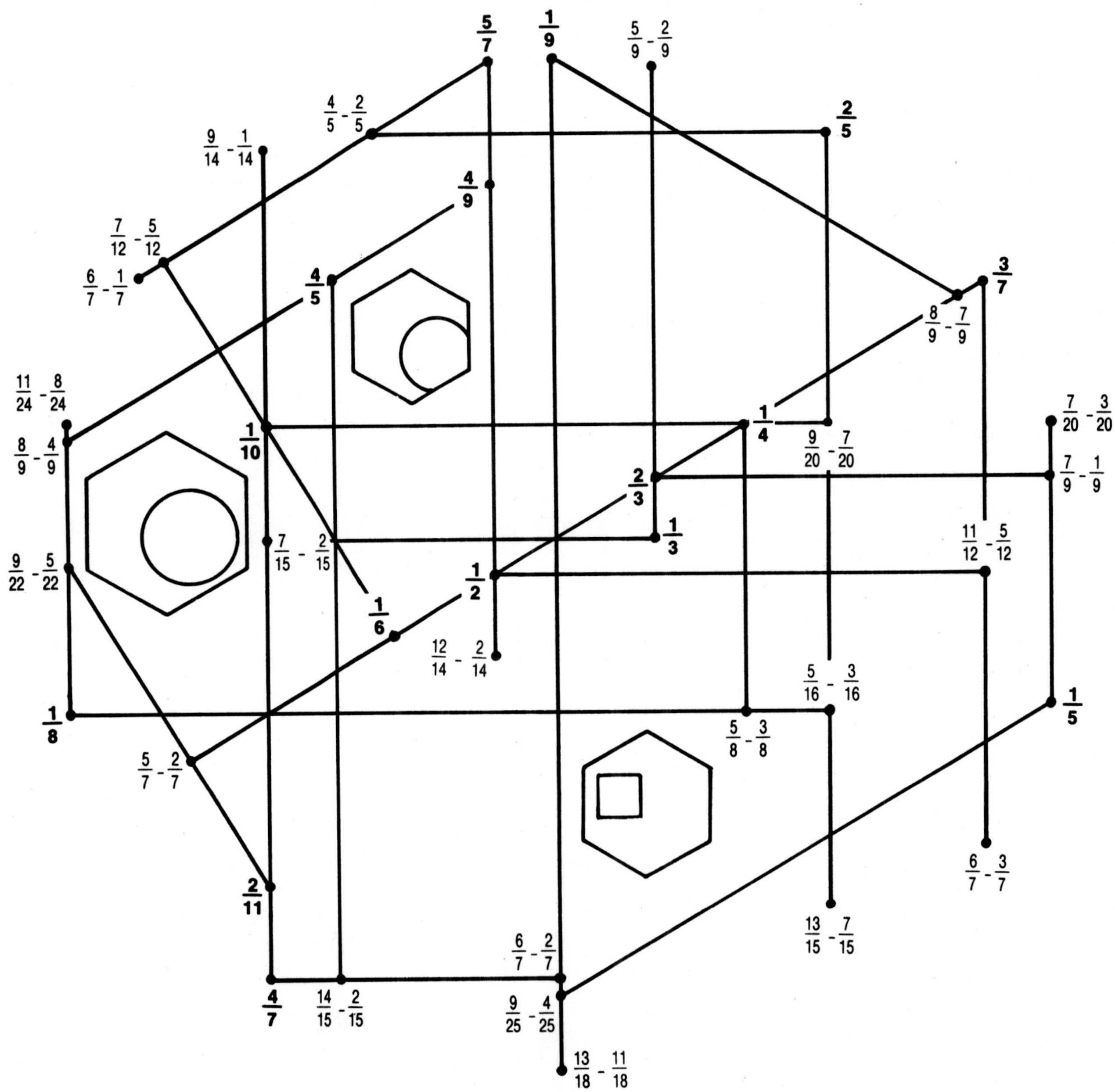

Name ______________________________ Date ________________

Read all of the directions before you begin.

1. **Choose one** of the **problems** next to a dot.
2. **Determine** the **difference** between the two fractions.
3. **Use a straight edge** and draw a line to the **correct answer**.
   Some answers may be used more than once.
4. **Make sure** you do **all** of the problems. Some lines pass through dots that are **not** part of the one being completed.
5. **You may** want to **color** or **shade your design** after you have completed the problems.

***NOTE:** The answers are **reduced** to their **simplest form**.

EXAMPLE: *Subtraction of fractional numbers with like denominators.* $\frac{5}{6} - \frac{2}{6} = \frac{5-2}{6} = \frac{3}{6} = \frac{1}{2}$

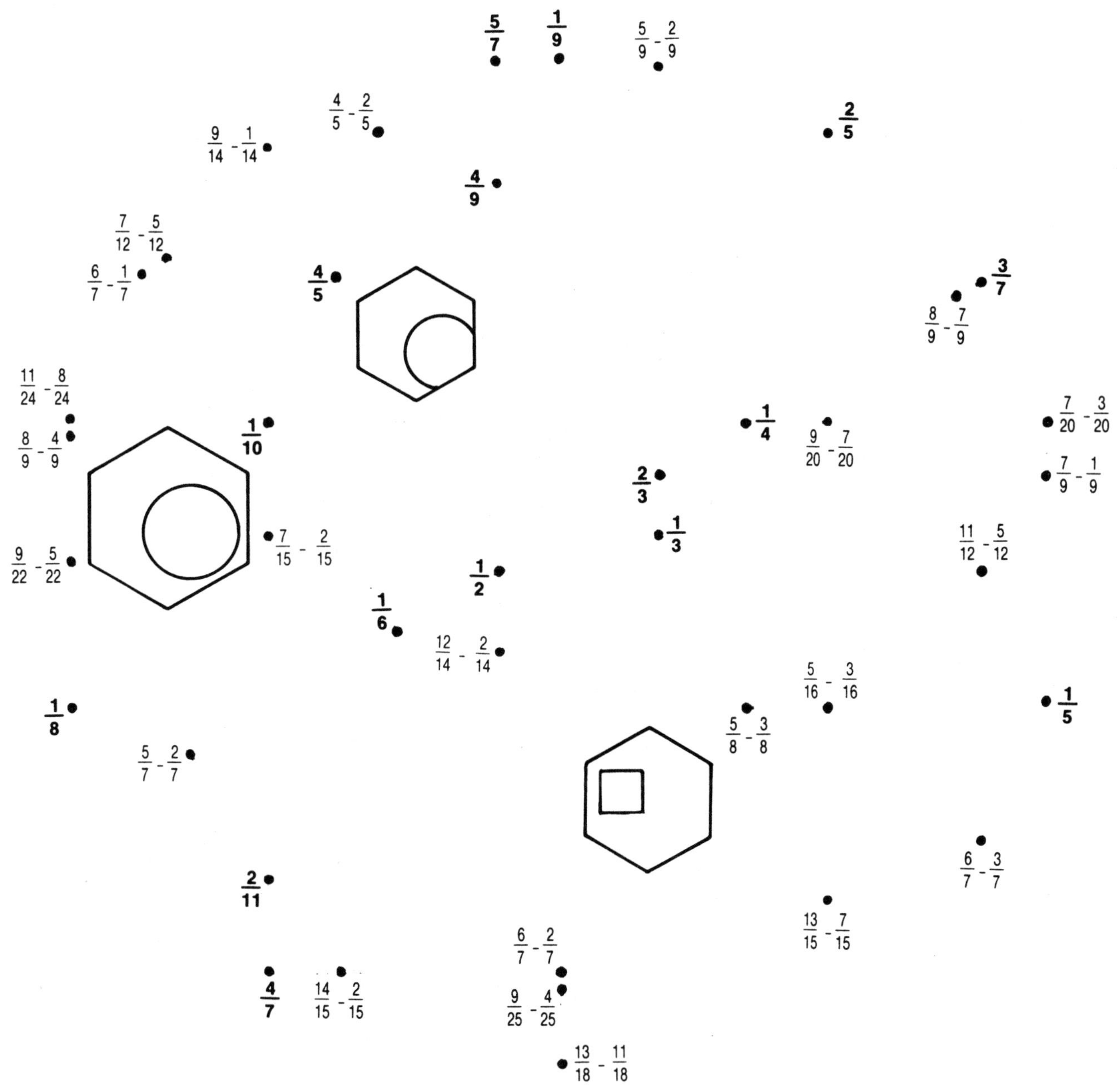

CORRECTION KEY

Designs in Math - Subtract Fractions ... Page 7

***Use this design as an answer key.**
***Allow your students to correct their own work.**
Make a transparency of this design and instruct your students to place the transparency over their completed design for a quick and easy check.

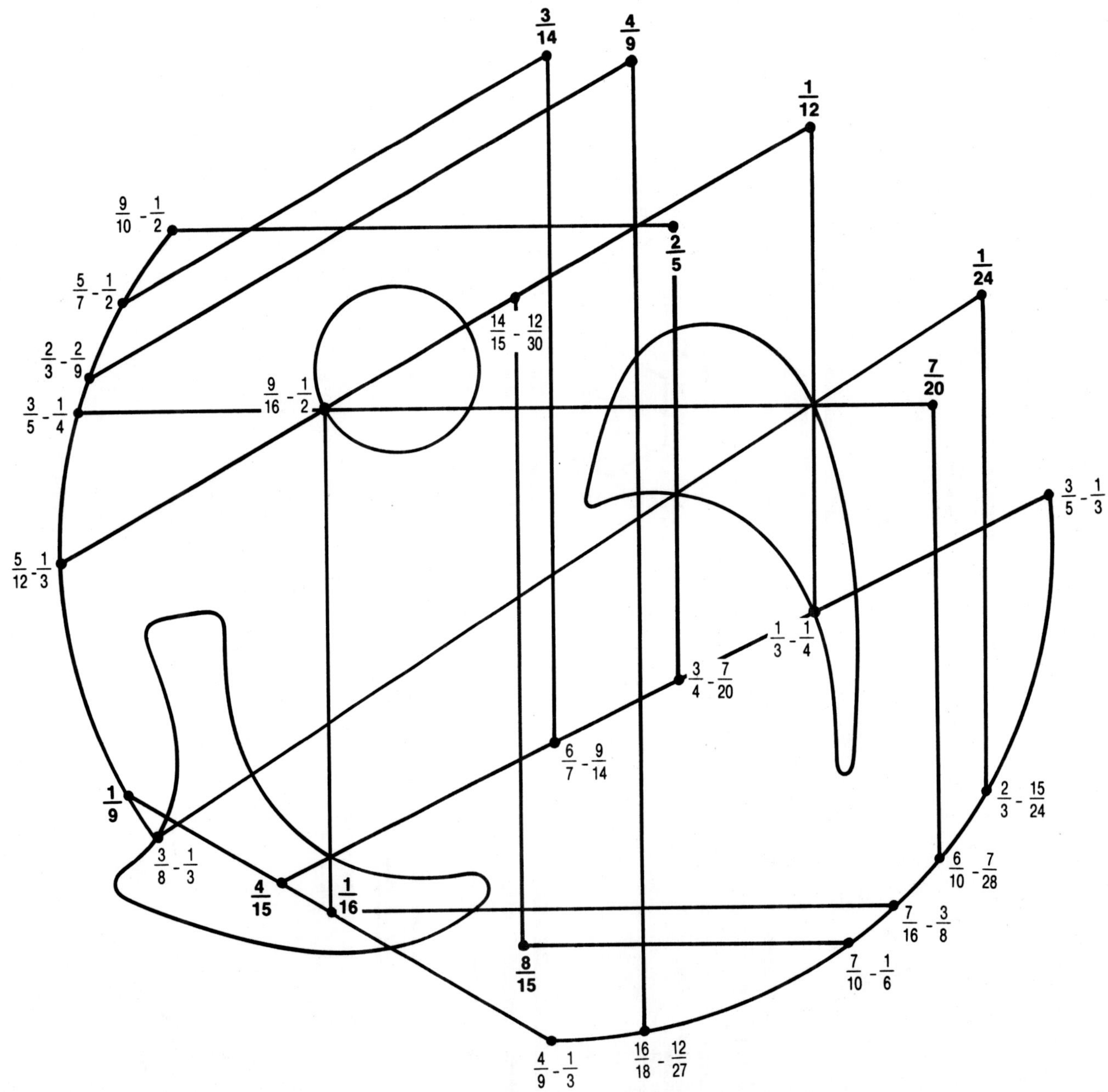

Name ______________________________ **Date** ______________

Read all of the directions before you begin.

1. **Choose one** of the **problems** next to a dot.
2. **Determine** the **difference** between the two fractions.
3. **Use a straight edge** and draw a line to the **correct answer**.
   Some answers may be used more than once.
4. **Make sure** you do **all** of the problems. Some lines pass through dots that are **not** part of the one being completed.
5. **You may** want to **color** or **shade your design** after you have completed the problems.

***NOTE:** The answers are **reduced** to their **simplest form**.

***EXAMPLE:** Subtraction of fractional numbers with unlike denominators.*

$$\frac{5}{6} - \frac{2}{12} =$$

$$\frac{10}{12} - \frac{2}{12} = \frac{10-2}{12} = \frac{8}{12} = \frac{2}{3}$$

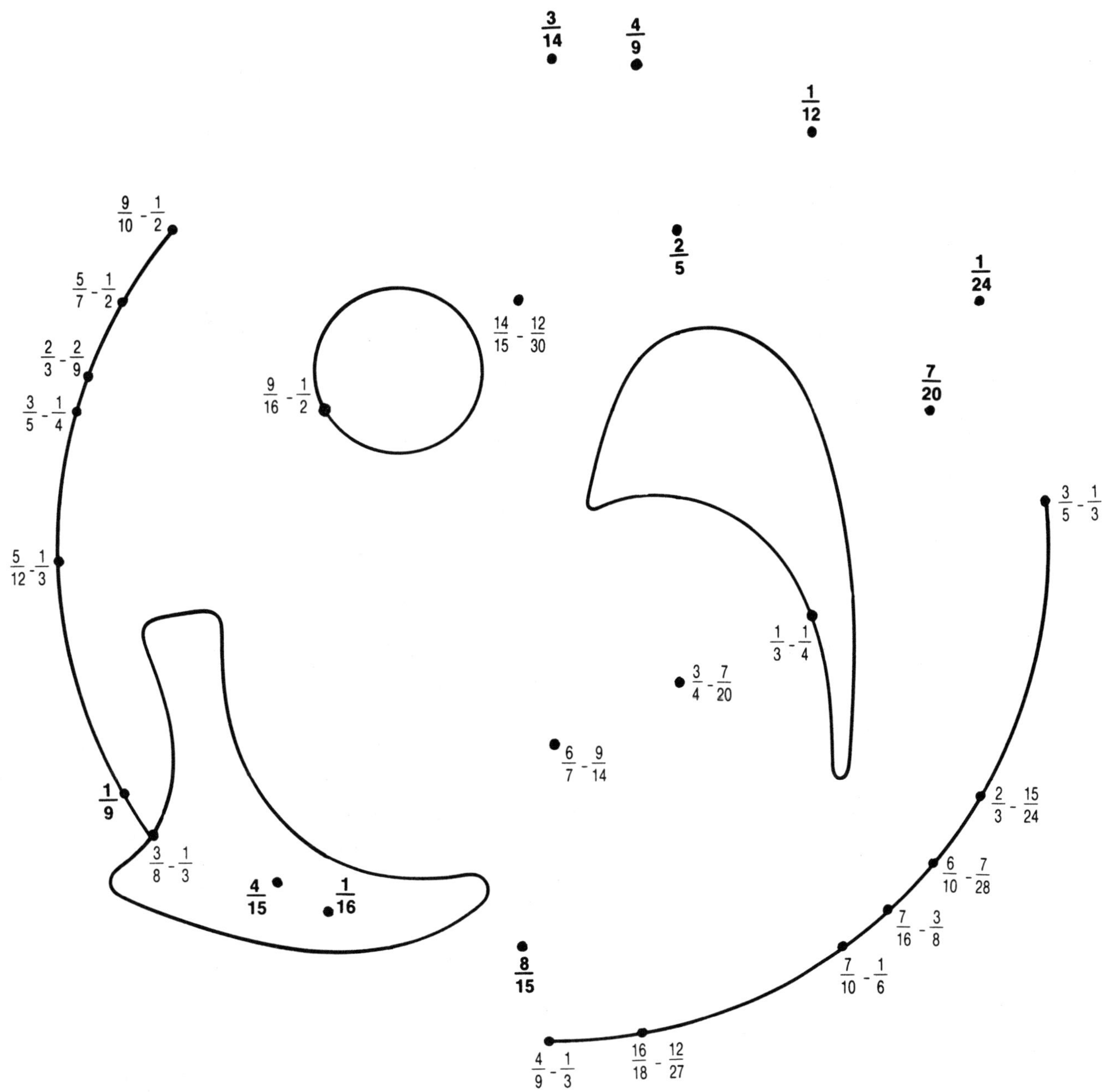

CORRECTION KEY

Designs in Math - Subtract Fractions... Page 8

***Use this design as an answer key.**

***Allow your students to correct their own work.**
Make a transparency of this design and instruct your students to place the transparency over their completed design for a quick and easy check.

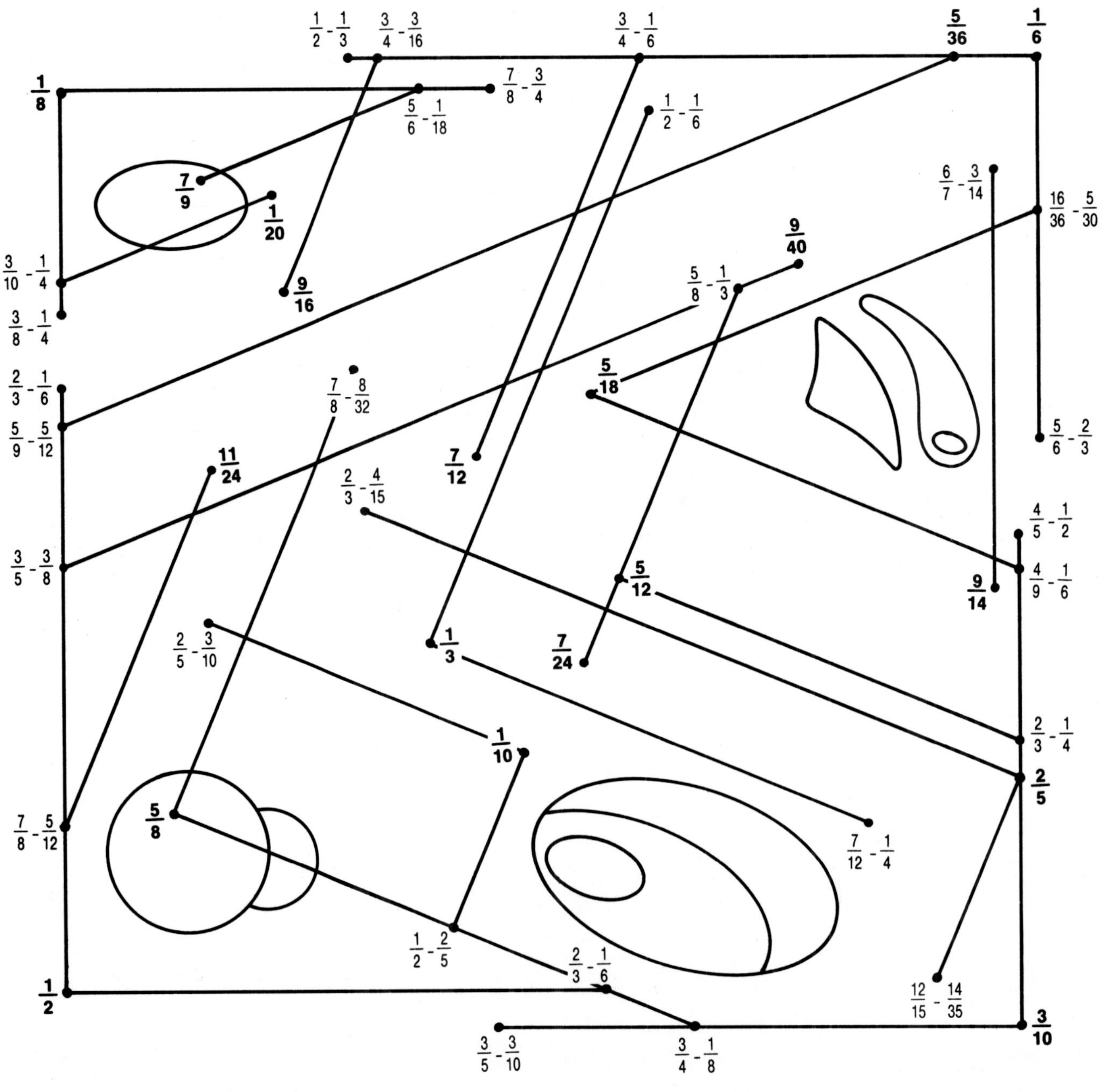

Name ______________________________ **Date** ______________

Read all of the directions before you begin.

1. **Choose one** of the **problems** next to a dot.
2. **Determine** the **difference** between the two fractions.
3. **Use a straight edge** and draw a line to the **correct answer**. Some answers may be used more than once.
4. **Make sure** you do **all** of the problems. Some lines pass through dots that are **not** part of the one being completed.
5. **You may** want to **color** or **shade your design** after you have completed the problems.

***NOTE:** The answers are **reduced** to their **simplest form**.

EXAMPLE: *Subtraction of fractional numbers with unlike denominators.*

$$\frac{5}{6} - \frac{2}{12} =$$

$$\frac{10}{12} - \frac{2}{12} = \frac{10-2}{12} = \frac{8}{12} = \frac{2}{3}$$

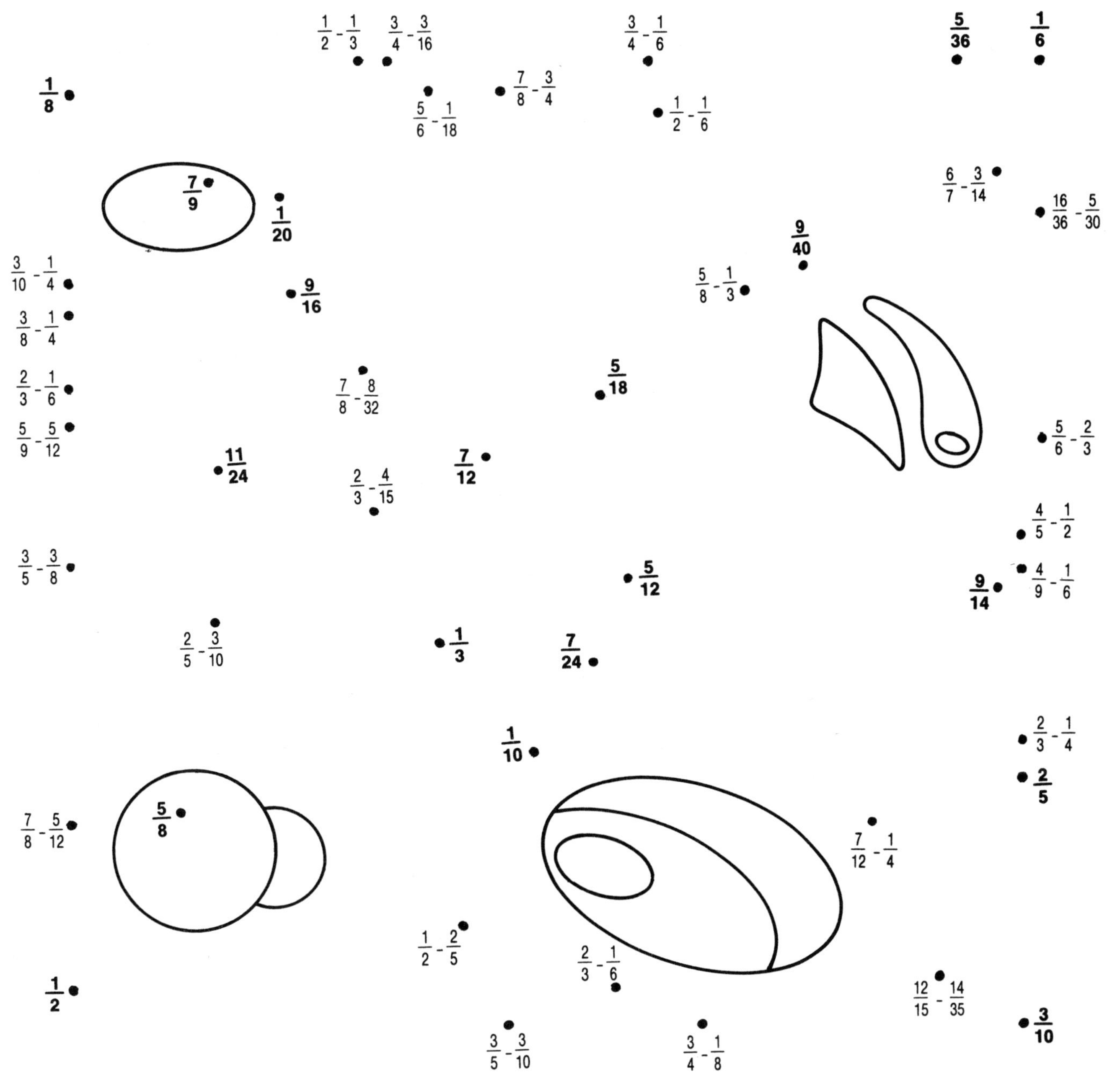

subtract fractions

CORRECTION KEY

Designs in Math - Subtract Fractions ... Page 9

***Use this design as an answer key.**
***Allow your students to correct their own work.**
Make a transparency of this design and instruct your students to place the transparency over their completed design for a quick and easy check.

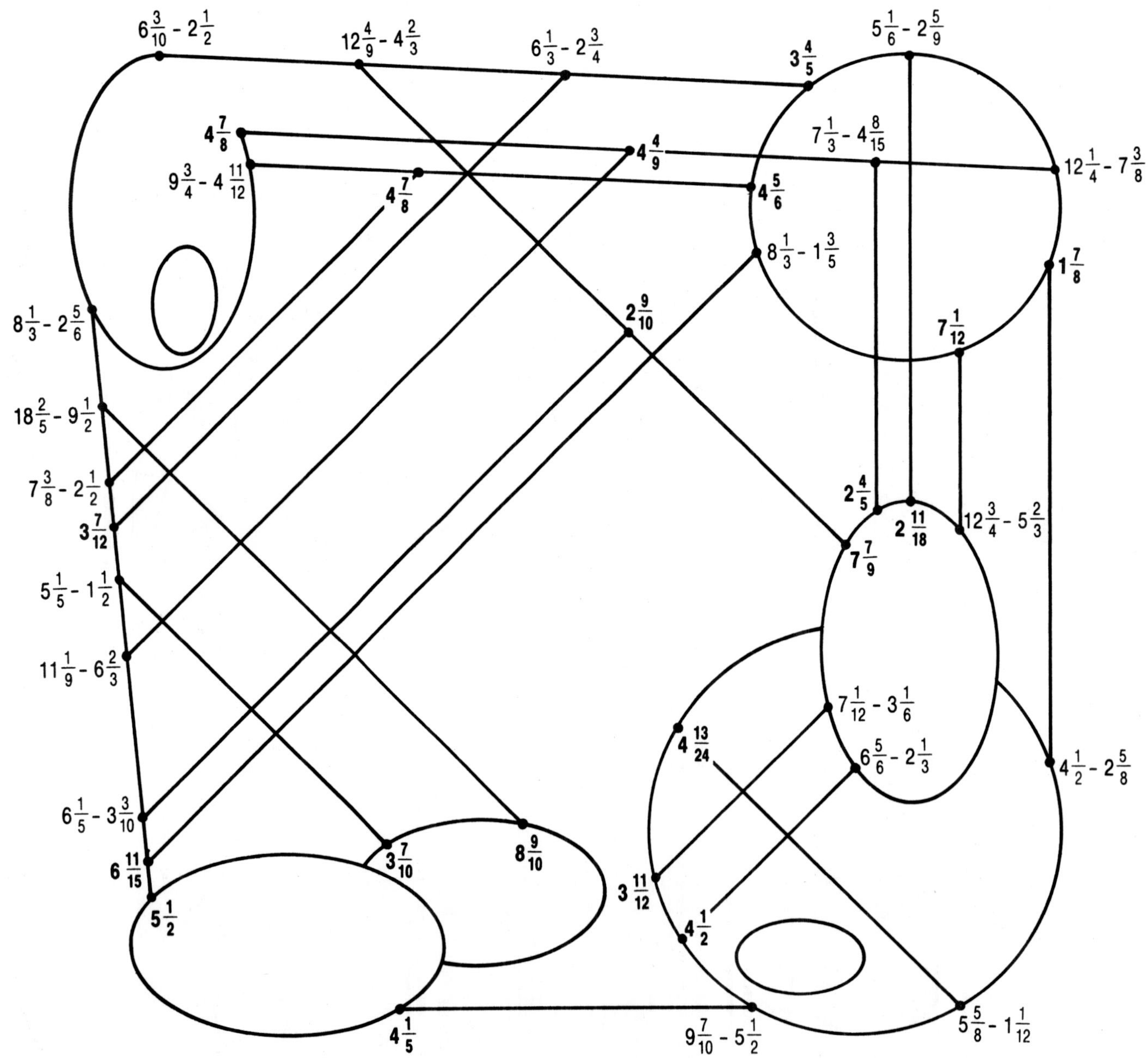

Name ______________________________ Date ______________

Read all of the directions before you begin.

1. **Choose one** of the **problems** next to a dot.
2. **Determine** the **difference** between the two fractions.
3. **Use a straight edge** and draw a line to the **correct answer**. Some answers may be used more than once.
4. **Make sure** you do **all** of the problems. Some lines pass through dots that are **not** part of the one being completed.
5. **You may** want to **color** or **shade your design** after you have completed the problems.

***NOTE:** The answers are **reduced** to their **simplest form**.

***EXAMPLE:** Subtraction of mixed numbers.* $6\frac{3}{10} - 2\frac{1}{2} = 6\frac{3}{10} - 2\frac{5}{10} = 5\frac{13}{10} - 2\frac{5}{10} = 3\frac{8}{10} = 3\frac{4}{5}$

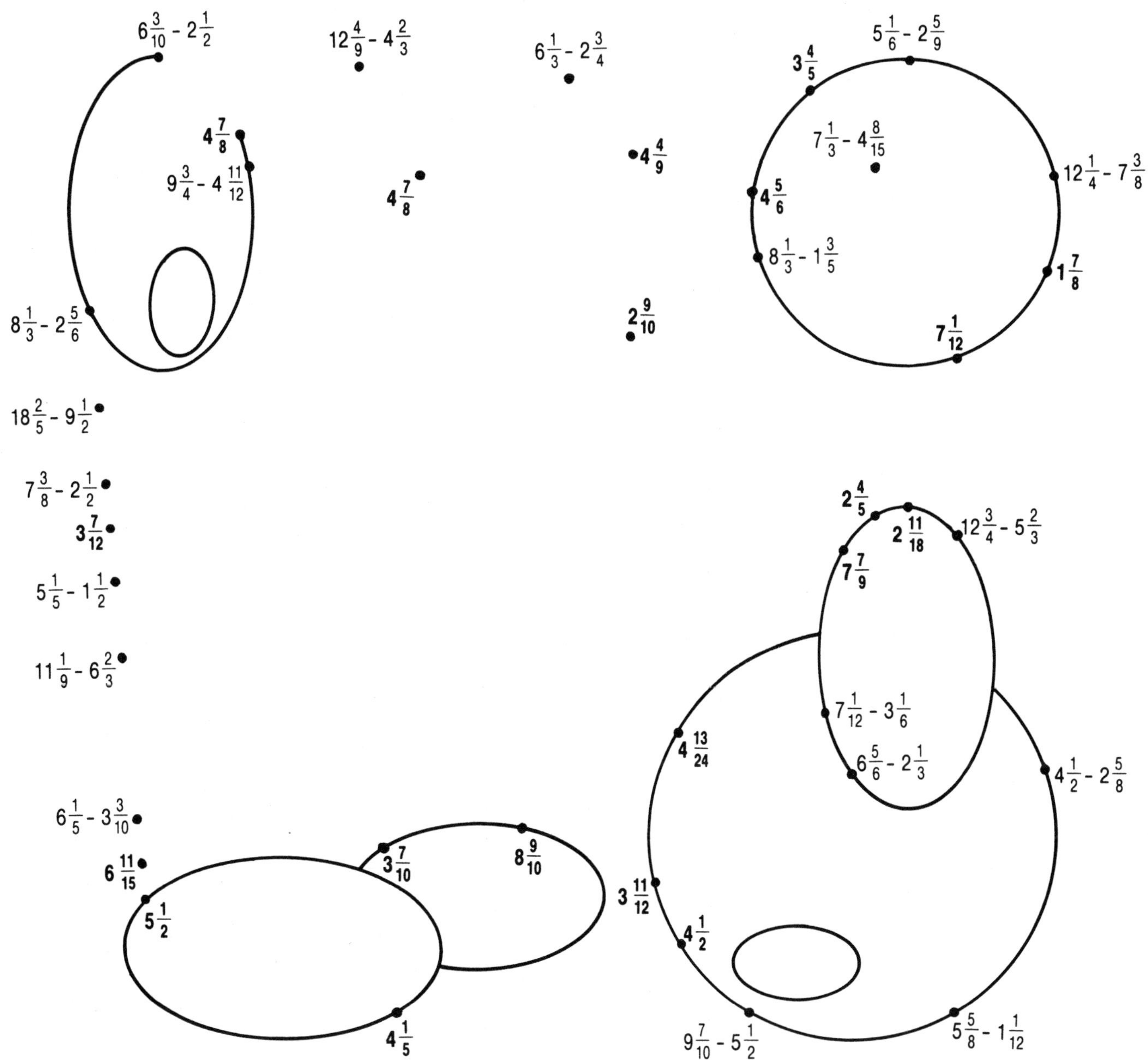

CORRECTION KEY

Designs in Math - Subtract Fractions ... Page 10

***Use this design as an answer key.**

***Allow your students to correct their own work.**
Make a transparency of this design and instruct your students to place the transparency over their completed design for a quick and easy check.

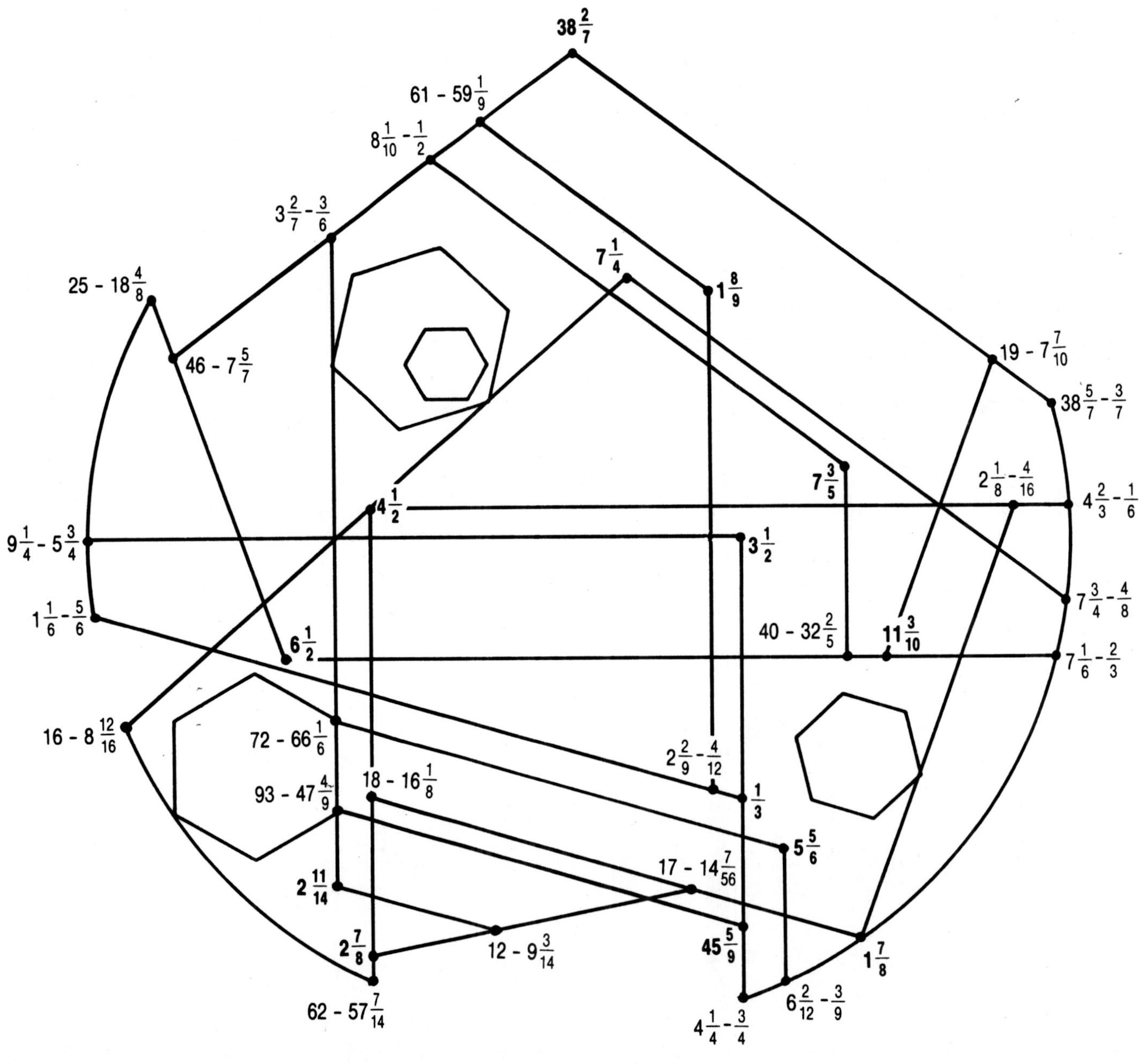

Name ____________________ **Date** ____________

Read all of the directions before you begin.

1. **Choose one** of the **problems** next to a dot.
2. **Determine** the **difference** between the two fractions.
3. **Use a straight edge** and draw a line to the **correct answer**. Some answers may be used more than once.
4. **Make sure** you do **all** of the problems. Some lines pass through dots that are **not** part of the one being completed.
5. **You may** want to **color** or **shade your design** after you have completed the problems.

***NOTE:** The answers are **reduced** to their **simplest form**.

***EXAMPLE:** Subtraction of whole numbers and mixed numbers.* $6 - 2\frac{3}{4} = 5\frac{4}{4} - 2\frac{3}{4} = 3\frac{1}{4}$

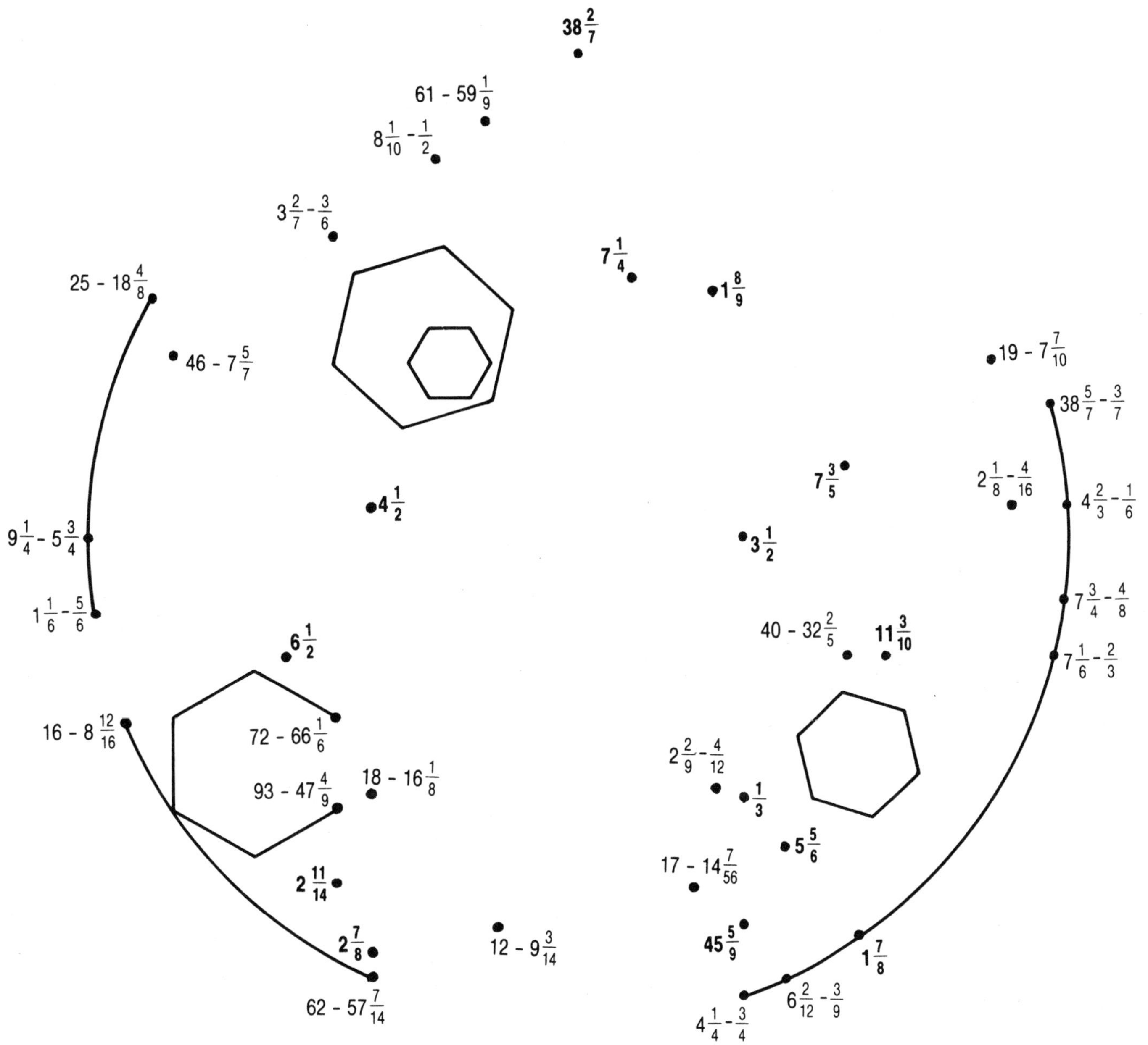

CORRECTION KEY

Designs in Math - Multiply Fractions ... Page 11

***Use this design as an answer key.**
***Allow your students to correct their own work.**
Make a transparency of this design and instruct your students to place the transparency over their completed design for a quick and easy check.

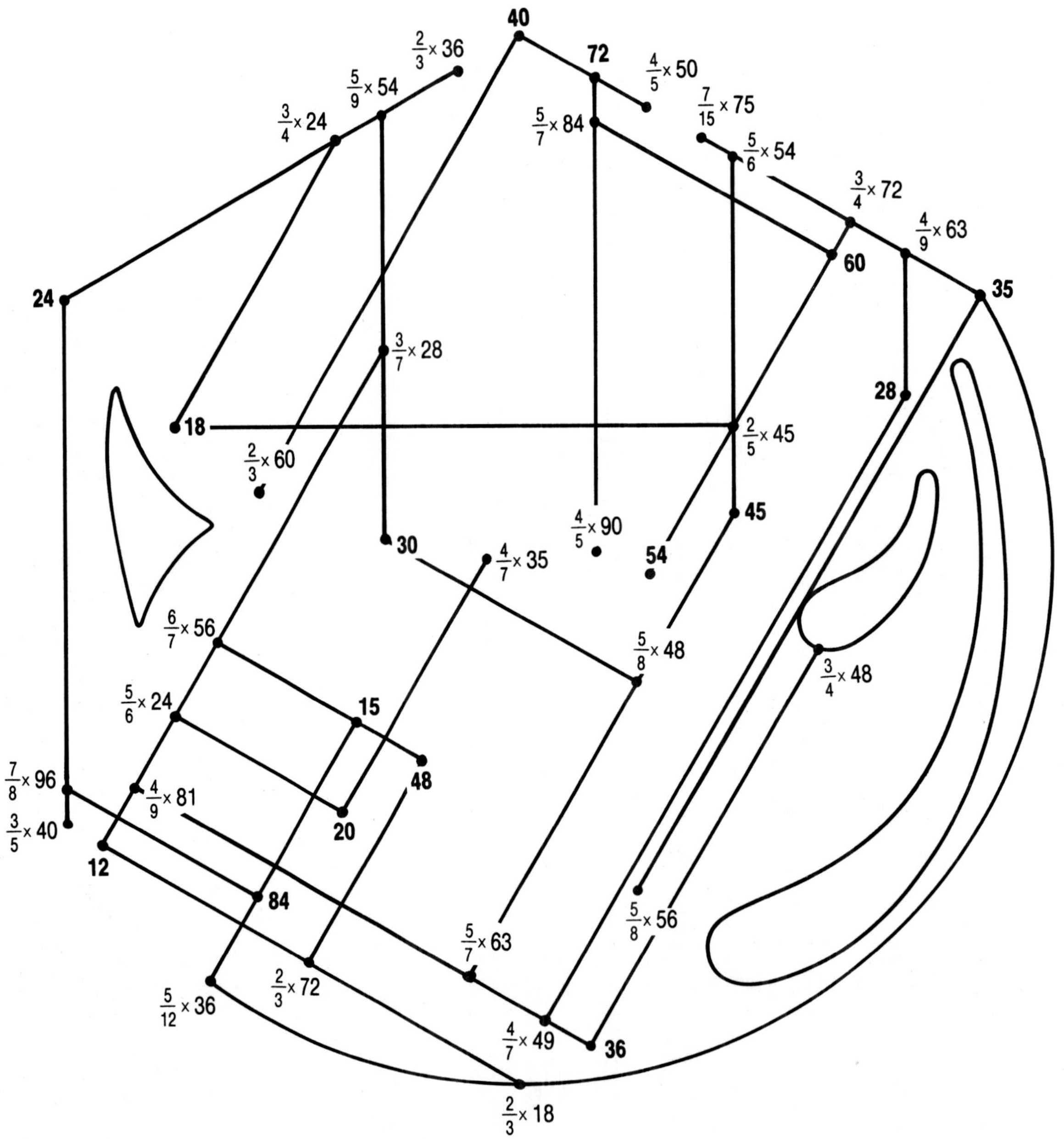

Name ______________________________ **Date** ________________

Read all of the directions before you begin.

1. **Choose one** of the **problems** next to a dot.
2. **Determine** the **product** of the fractional number and the whole number.
3. **Use a straight edge** and draw a line to the **correct answer**. Some answers may be used more than once.
4. **Make sure** you do **all** of the problems. Some lines pass through dots that are **not** part of the one being completed.
5. **You may** want to **color** or **shade your design** after you have completed the problems.

EXAMPLE: *Multiplication of fractional numbers and whole numbers.* $\frac{3}{4} \times 16 = \frac{3}{\not{4}_{1}} \times \frac{\not{16}^{4}}{1} = \frac{12}{1} = 12$

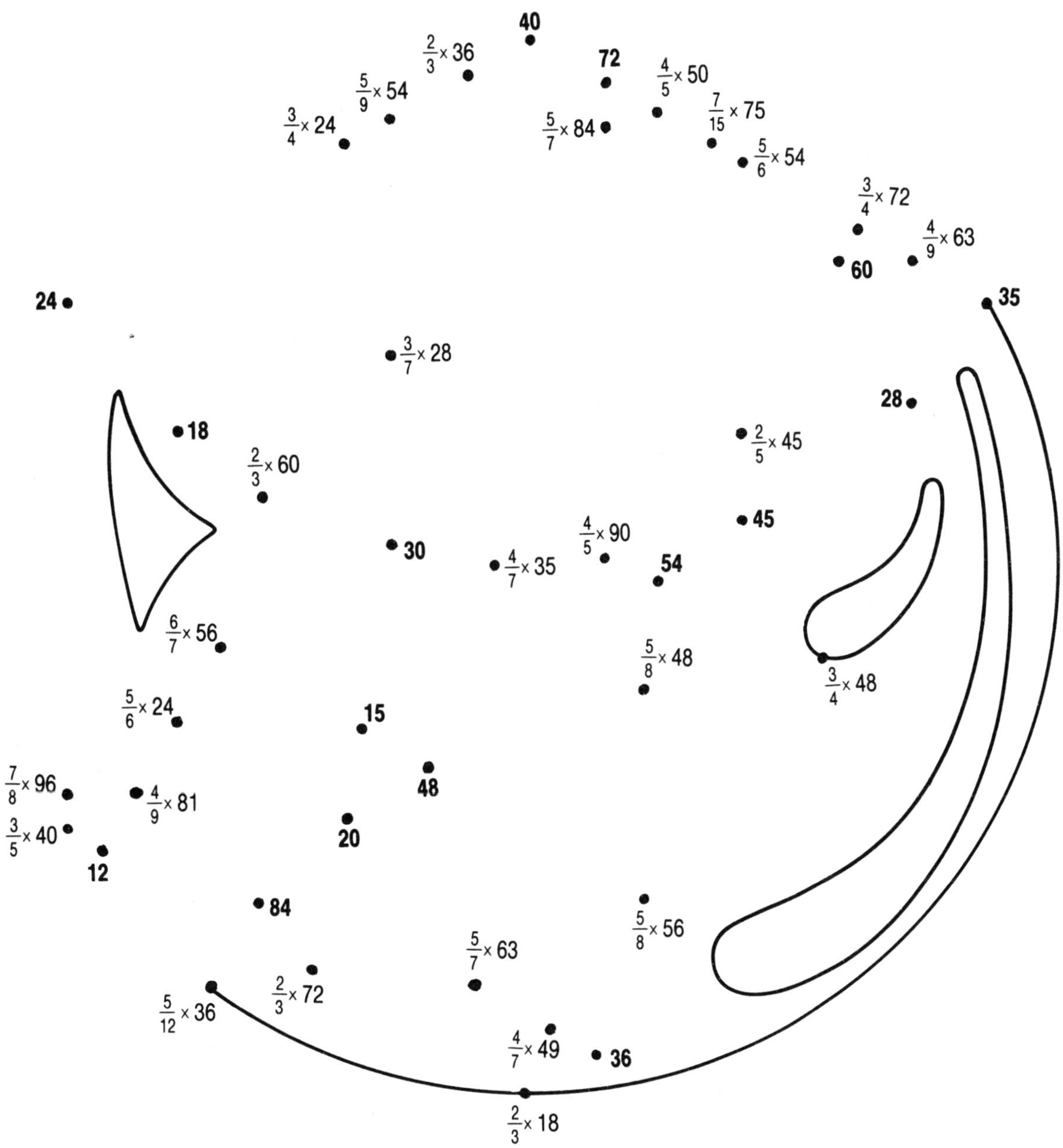

CORRECTION KEY

Designs in Math - Multiply Fractions ... Page 12

***Use this design as an answer key.**
***Allow your students to correct their own work.**
Make a transparency of this design and instruct your students to place the transparency over their completed design for a quick and easy check.

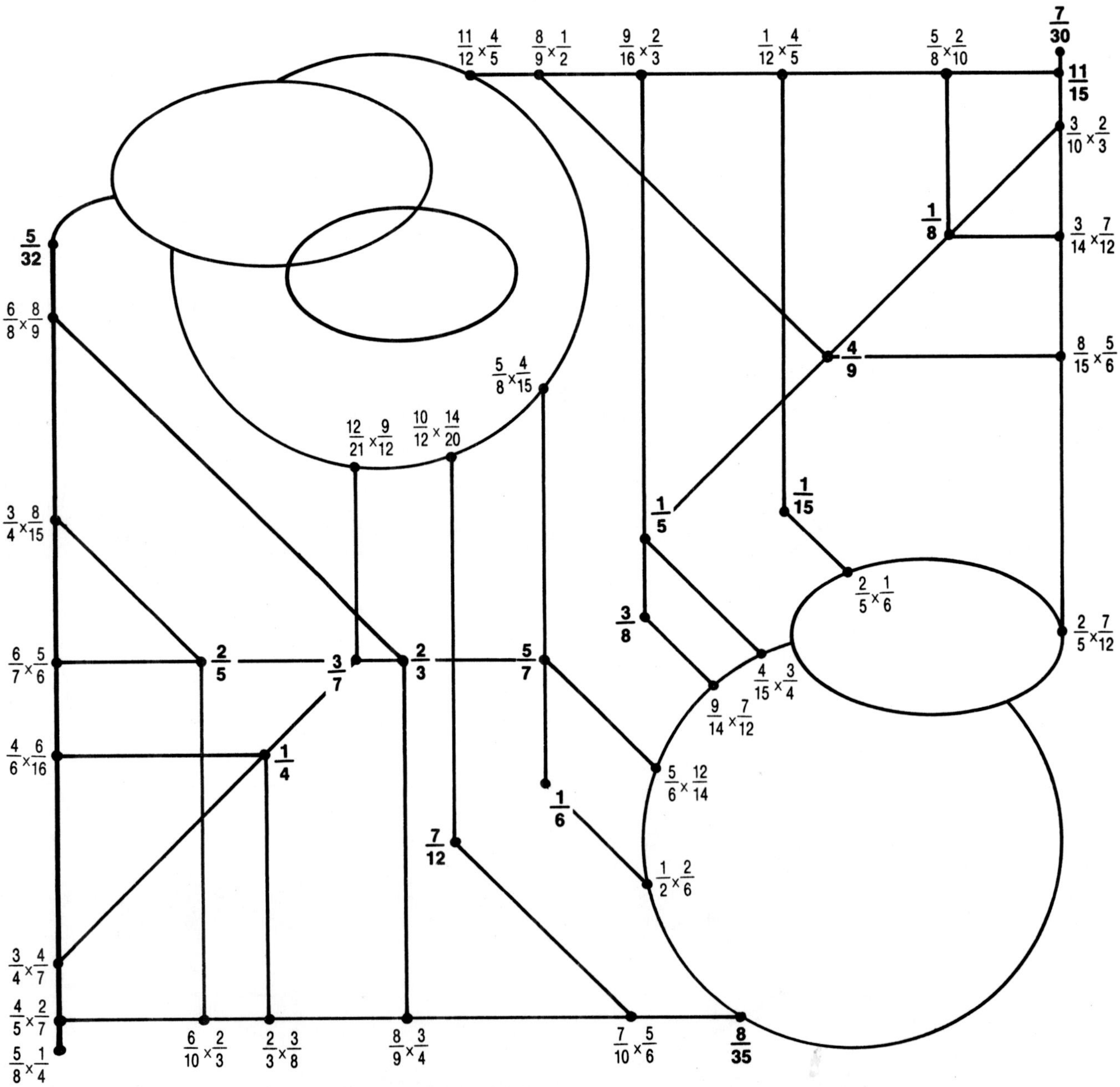

Name ______________________ **Date** ______________

Read all of the directions before you begin.

1. **Choose one** of the **problems** next to a dot.
2. **Determine** the **product** of the two fractions.
3. **Use a straight edge** and draw a line to the **correct answer**. Some answers may be used more than once.
4. **Make sure** you do **all** of the problems. Some lines pass through dots that are **not** part of the one being completed.
5. **You may** want to **color** or **shade your design** after you have completed the problems.

***NOTE:** The answers are **reduced** to their **simplest form**.

EXAMPLE: *Multiplication of fractional numbers.* $\frac{\not{2}^{1}}{\not{3}_{1}} \times \frac{\not{3}^{1}}{\not{8}_{4}} = \frac{1}{4}$

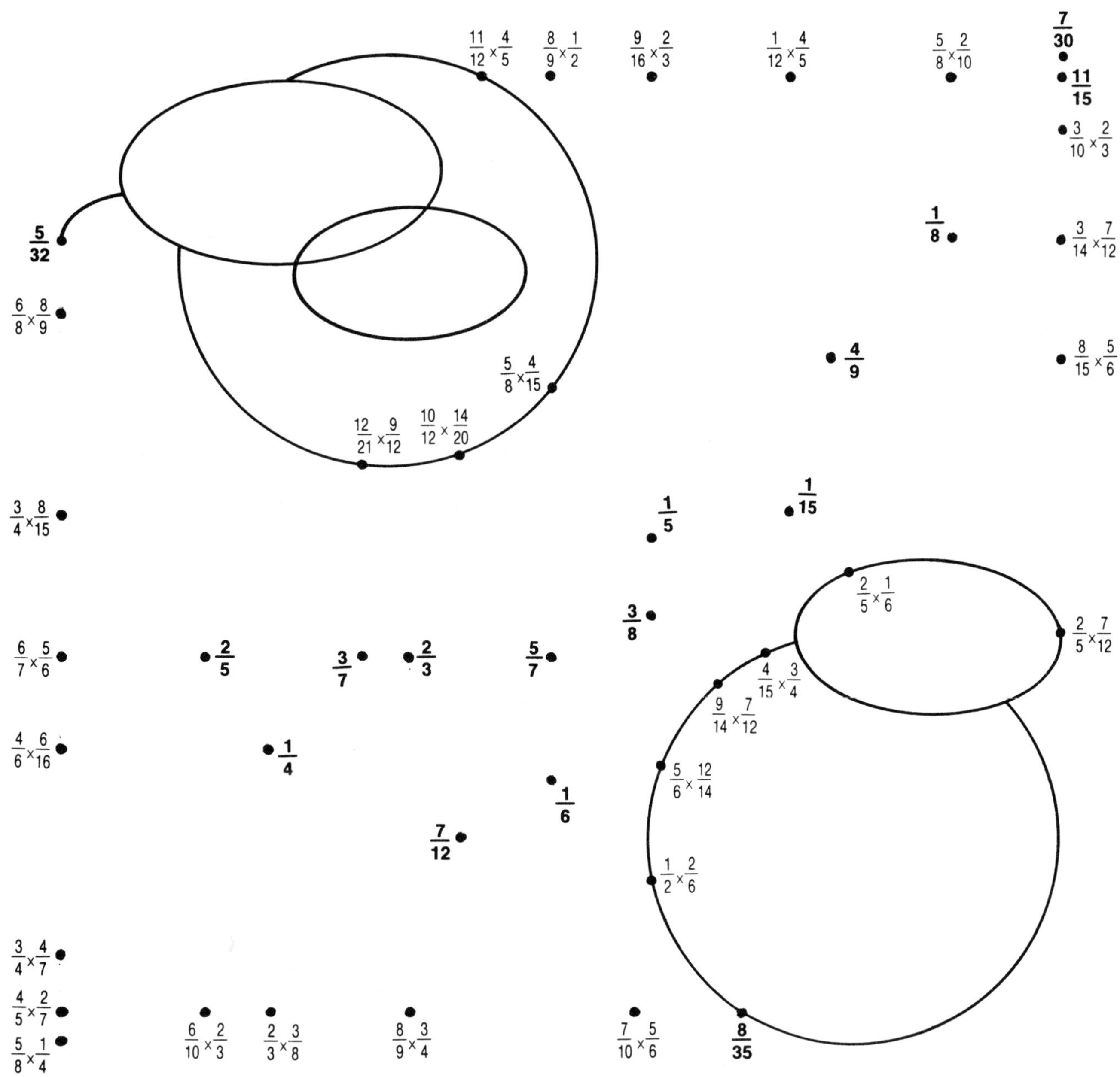

CORRECTION KEY

Designs in Math - Multiply Fractions ... Page 13

***Use this design as an answer key.**

***Allow your students to correct their own work.**
Make a transparency of this design and instruct your students to place the transparency over their completed design for a quick and easy check.

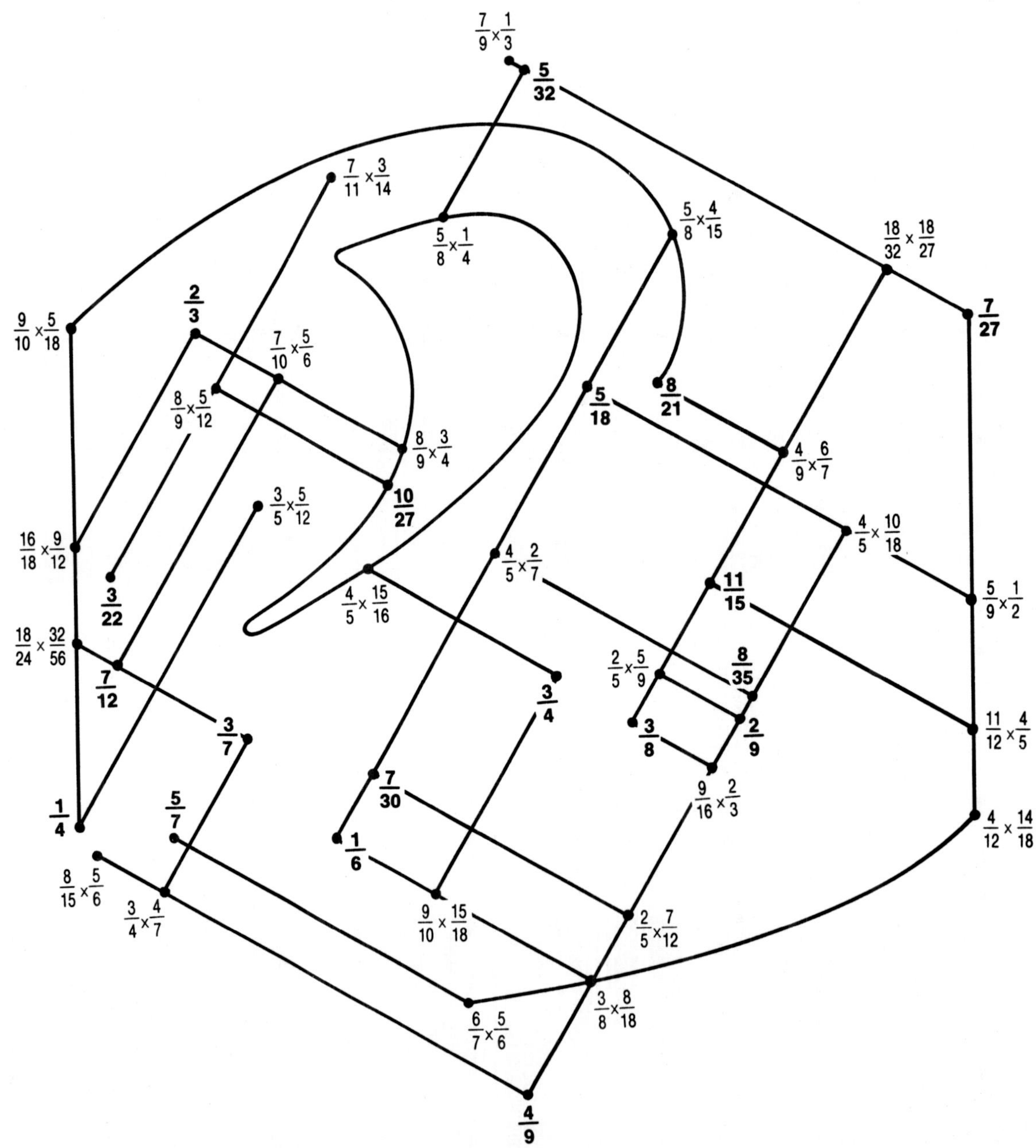

Name ______________________________ **Date** ________________

Read all of the directions before you begin.

1. **Choose one** of the **problems** next to a dot.
2. **Determine** the **product** of the two fractions.
3. **Use a straight edge** and draw a line to the **correct answer**.
   Some answers may be used more than once.
4. **Make sure** you do **all** of the problems. Some lines pass through dots that are **not** part of the one being completed.
5. **You may** want to **color** or **shade your design** after you have completed the problems.

***NOTE:** The answers are **reduced** to their **simplest form**.

EXAMPLE: *Multiplication of fractional numbers.* $\frac{\overset{1}{\cancel{2}}}{\underset{1}{\cancel{3}}} \times \frac{\overset{1}{\cancel{3}}}{\underset{4}{\cancel{8}}} = \frac{1}{4}$

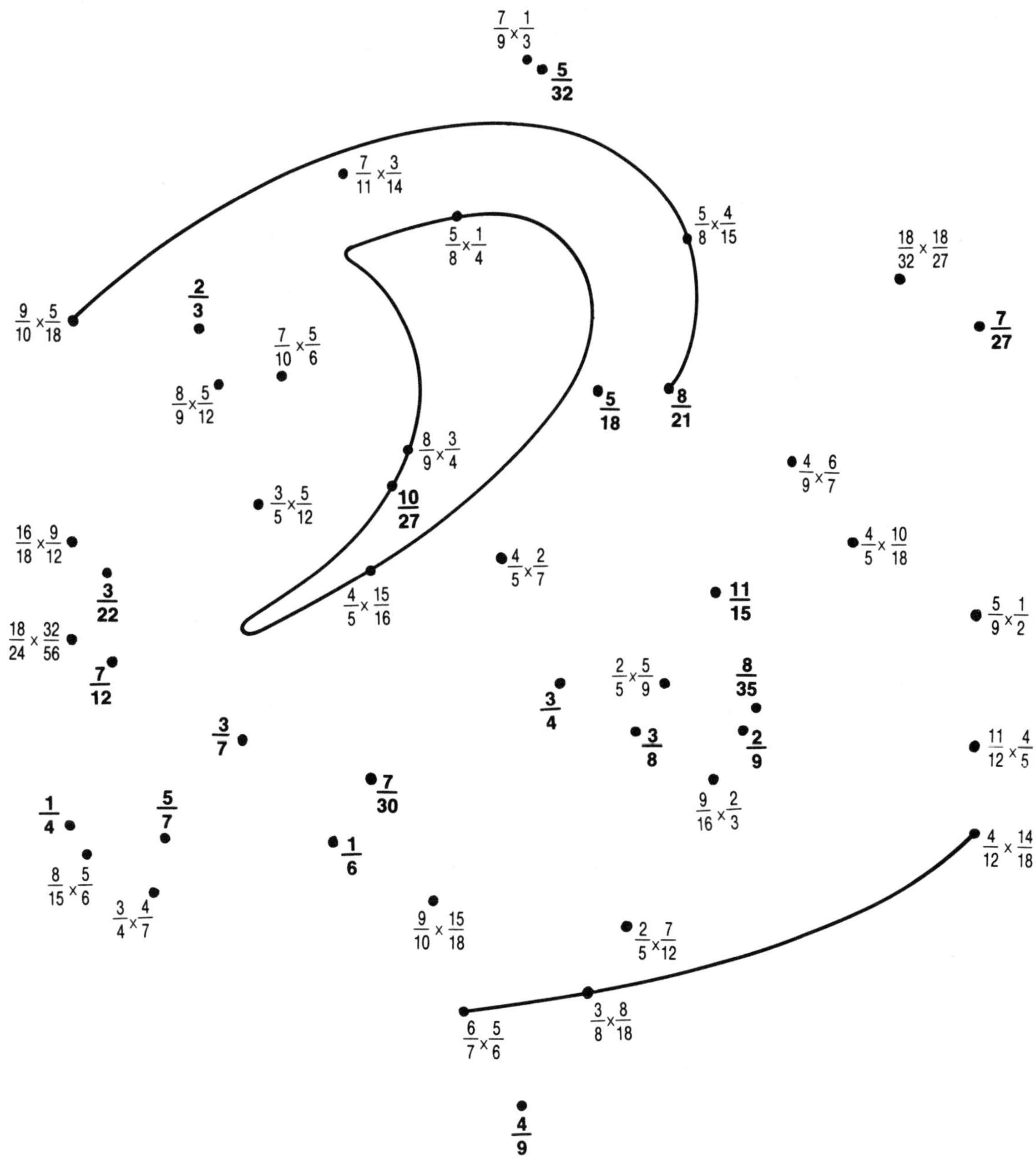

CORRECTION KEY

Designs in Math - Multiply Fractions ... Page 14

***Use this design as an answer key.**
***Allow your students to correct their own work.**
Make a transparency of this design and instruct your students to place the transparency over their completed design for a quick and easy check.

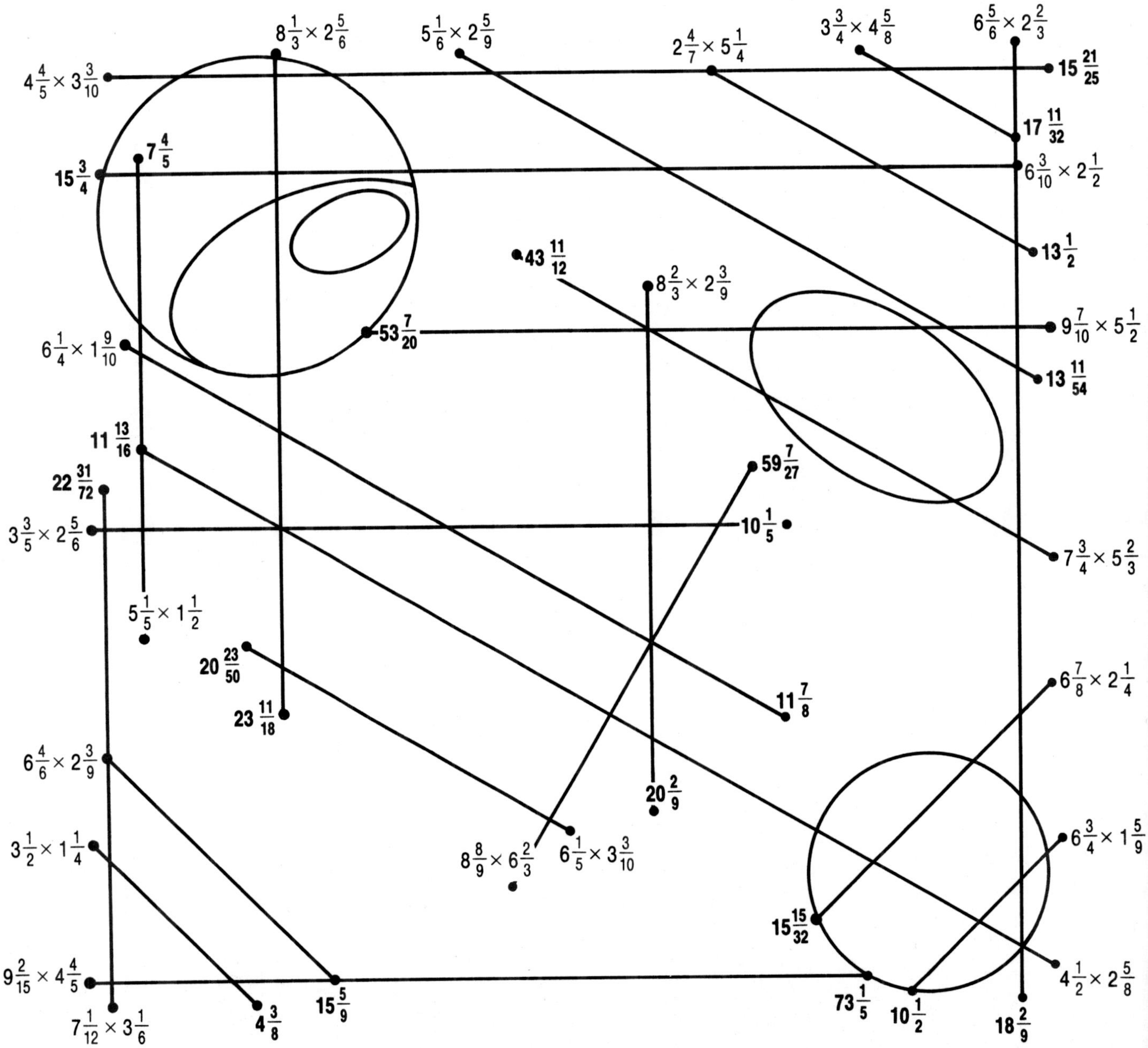

Name ______________________________ Date ______________

Read all of the directions before you begin.

1. **Choose one** of the **problems** next to a dot.
2. **Determine** the **product** of the two fractions.
3. **Use a straight edge** and draw a line to the **correct answer**.
   Some answers may be used more than once.
4. **Make sure** you do **all** of the problems. Some lines pass through dots that are **not** part of the one being completed.
5. **You may** want to **color** or **shade your design** after you have completed the problems.

***NOTE:** The answers are **reduced** to their **simplest form**.

***EXAMPLE:** Multiplication of mixed numbers.*

$$3\frac{8}{9} \times 1\frac{5}{7} = \frac{\overset{5}{\cancel{35}}}{\underset{3}{\cancel{9}}} \times \frac{\overset{4}{\cancel{12}}}{\underset{1}{\cancel{7}}} = \frac{20}{3} = 6\frac{2}{3}$$

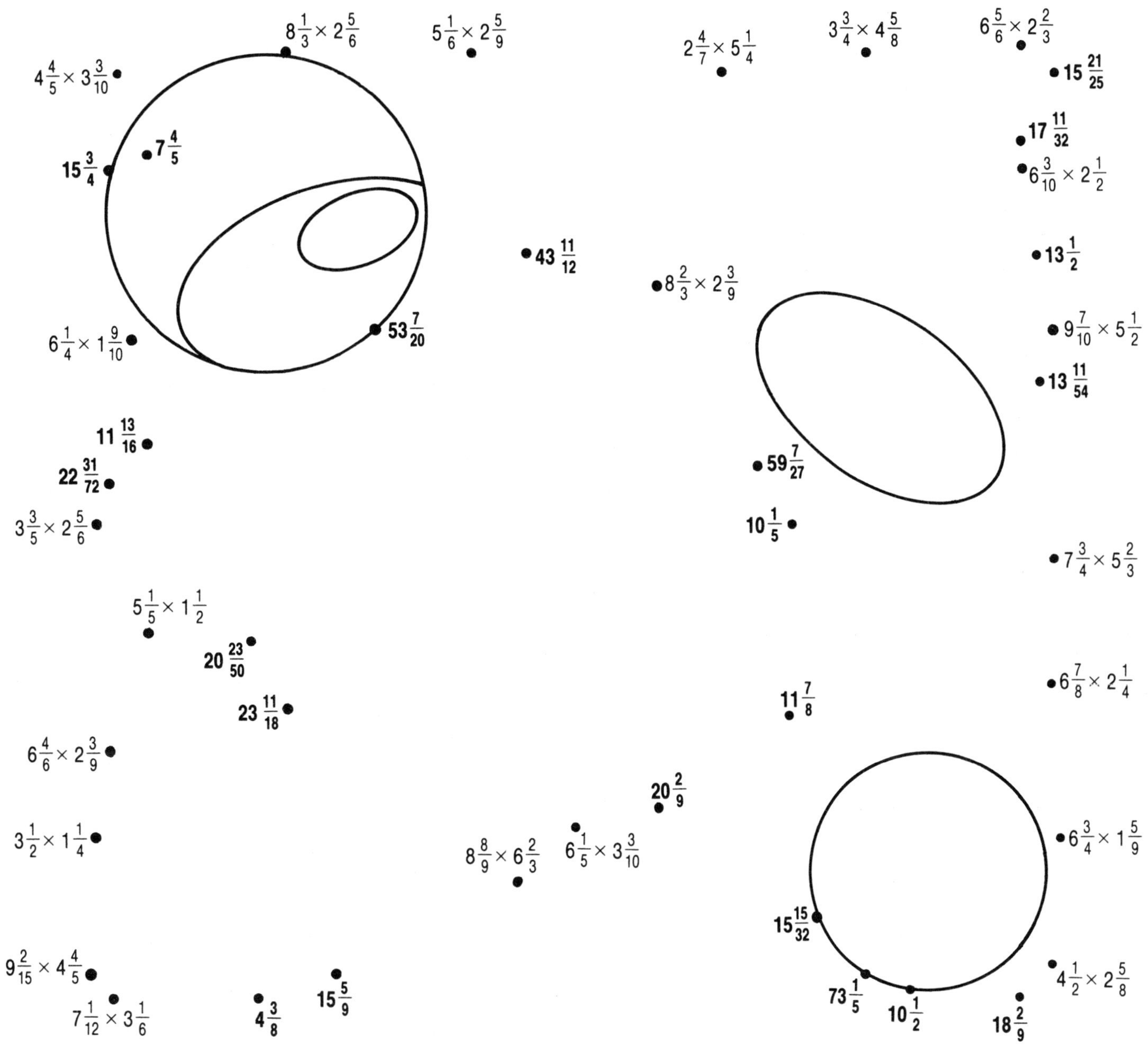

CORRECTION KEY

Designs in Math - Multiply Fractions ... Page 15

***Use this design as an answer key.**
***Allow your students to correct their own work.**
Make a transparency of this design and instruct your students to place the transparency over their completed design for a quick and easy check.

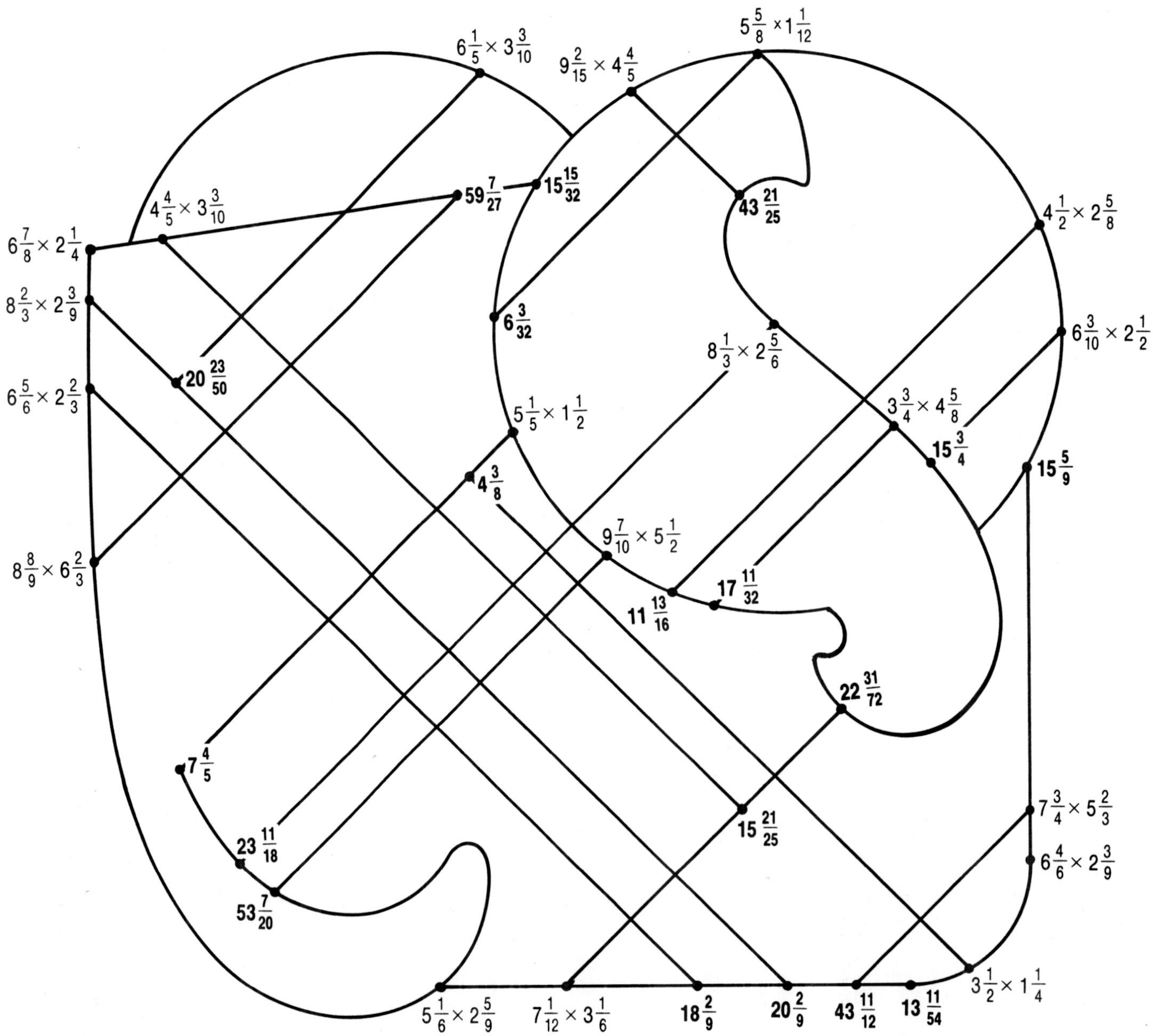

Name ______________________________ **Date** ______________

Read all of the directions before you begin.

1. **Choose one** of the **problems** next to a dot.
2. **Determine** the **product** of the two fractions.
3. **Use a straight edge** and draw a line to the **correct answer**. Some answers may be used more than once.
4. **Make sure** you do **all** of the problems. Some lines pass through dots that are **not** part of the one being completed.
5. **You may** want to **color** or **shade your design** after you have completed the problems.

***NOTE:** The answers are **reduced** to their **simplest form**.

EXAMPLE: *Multiplication of mixed numbers.*

$$3\frac{8}{9} \times 1\frac{5}{7} = \frac{\overset{5}{\cancel{35}}}{\underset{3}{\cancel{9}}} \times \frac{\overset{4}{\cancel{12}}}{\underset{1}{\cancel{7}}} = \frac{20}{3} = 6\frac{2}{3}$$

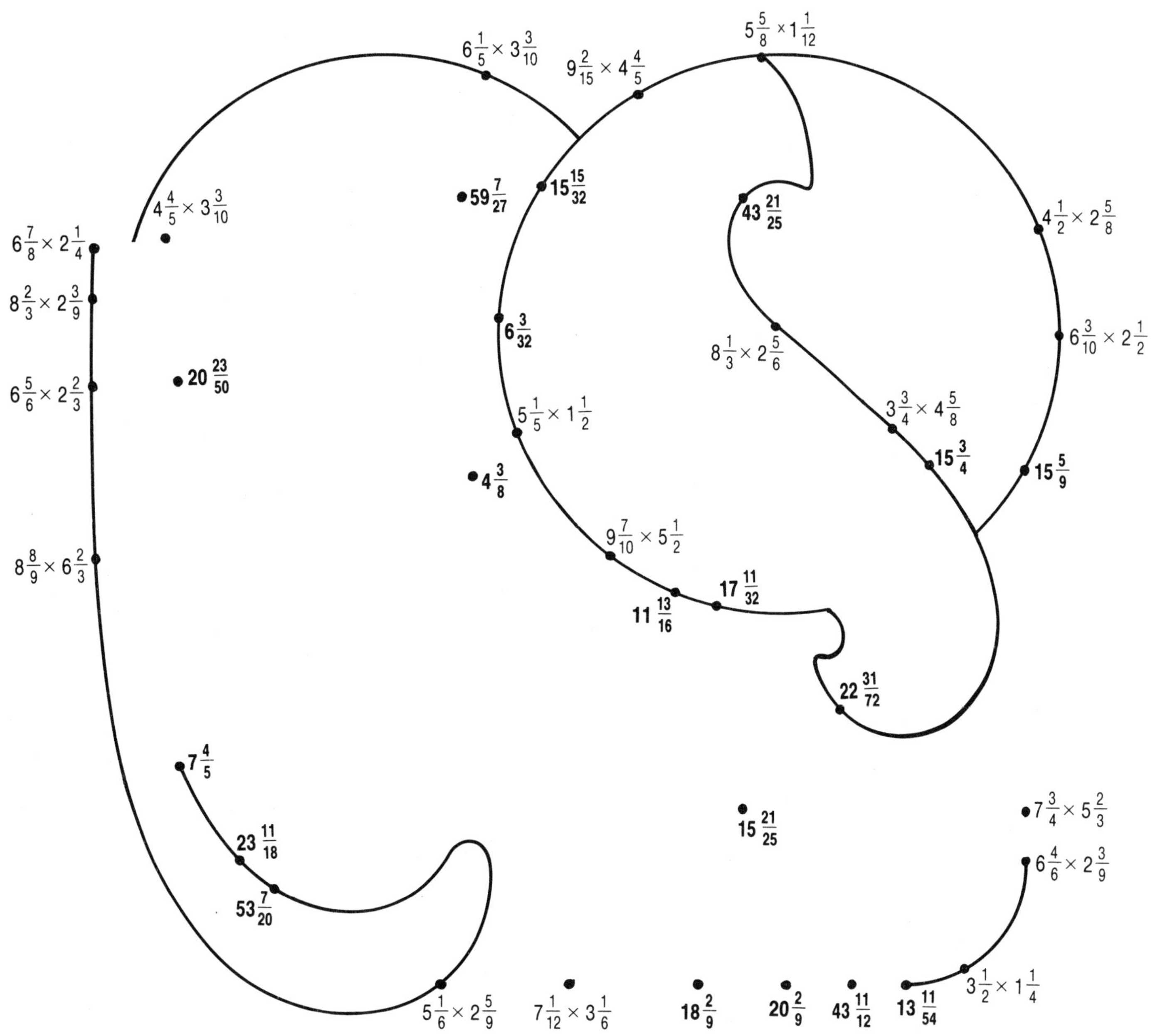

CORRECTION KEY

Designs in Math - Divide Fractions ... Page 16

***Use this design as an answer key.**
***Allow your students to correct their own work.**
Make a transparency of this design and instruct your students to place the transparency over their completed design for a quick and easy check.

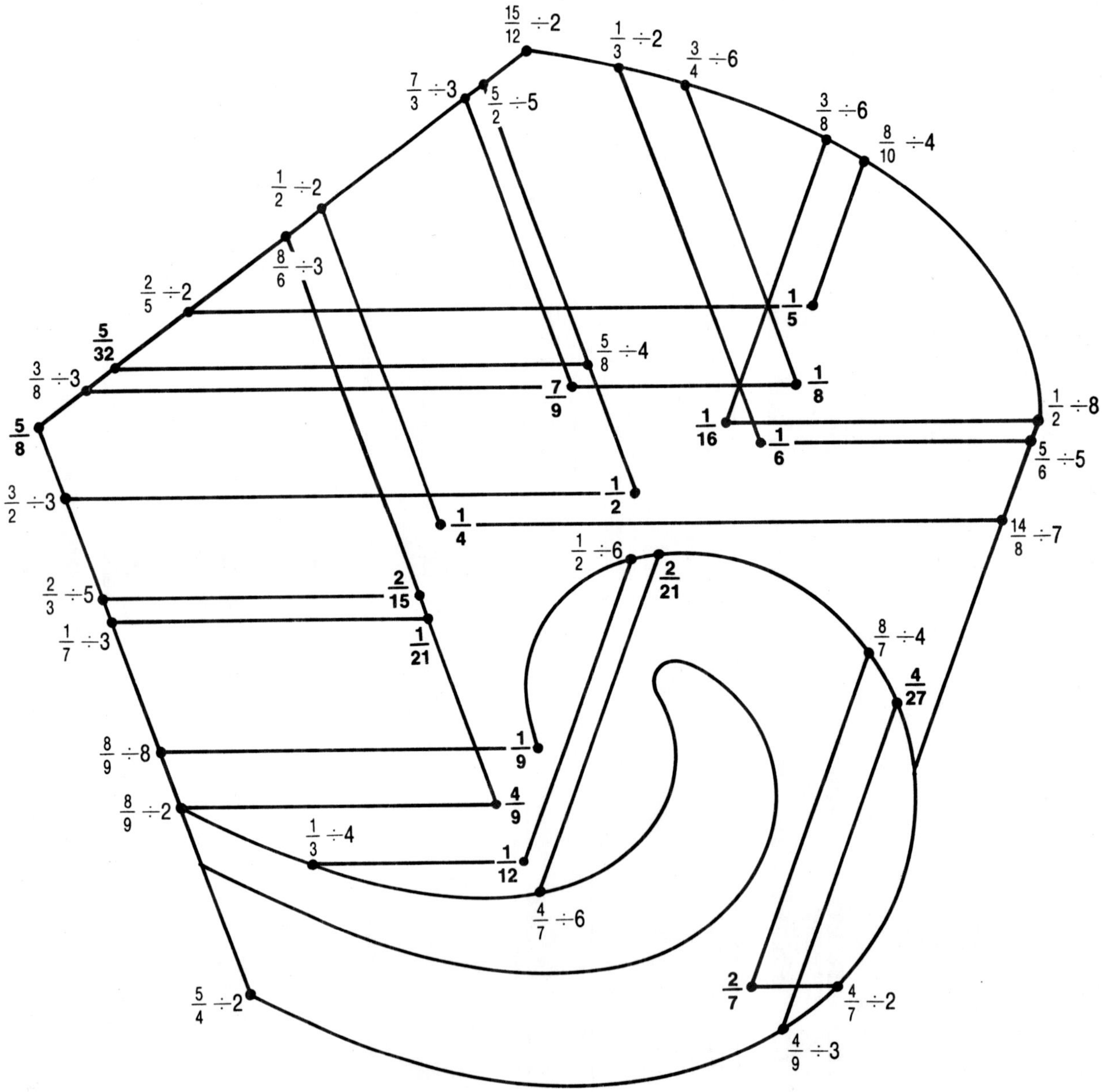

Name ______________________________ Date ________________

Read all of the directions before you begin.

1. **Choose one** of the **problems** next to a dot.
2. **Determine** the **quotient** of the fractional number and the whole number.
3. **Use a straight edge** and draw a line to the **correct answer**.
   Some answers may be used more than once.
4. **Make sure** you do **all** of the problems. Some lines pass through dots that are **not** part of the one being completed.
5. **You may** want to **color** or **shade your design** after you have completed the problems.

***NOTE:** The answers are **reduced** to their **simplest form**.

EXAMPLE: *Division of fractional numbers and whole numbers.* $\frac{3}{4} \div 9 = \frac{\overset{1}{\cancel{3}}}{4} \times \frac{1}{\underset{3}{\cancel{9}}} = \frac{1}{12}$

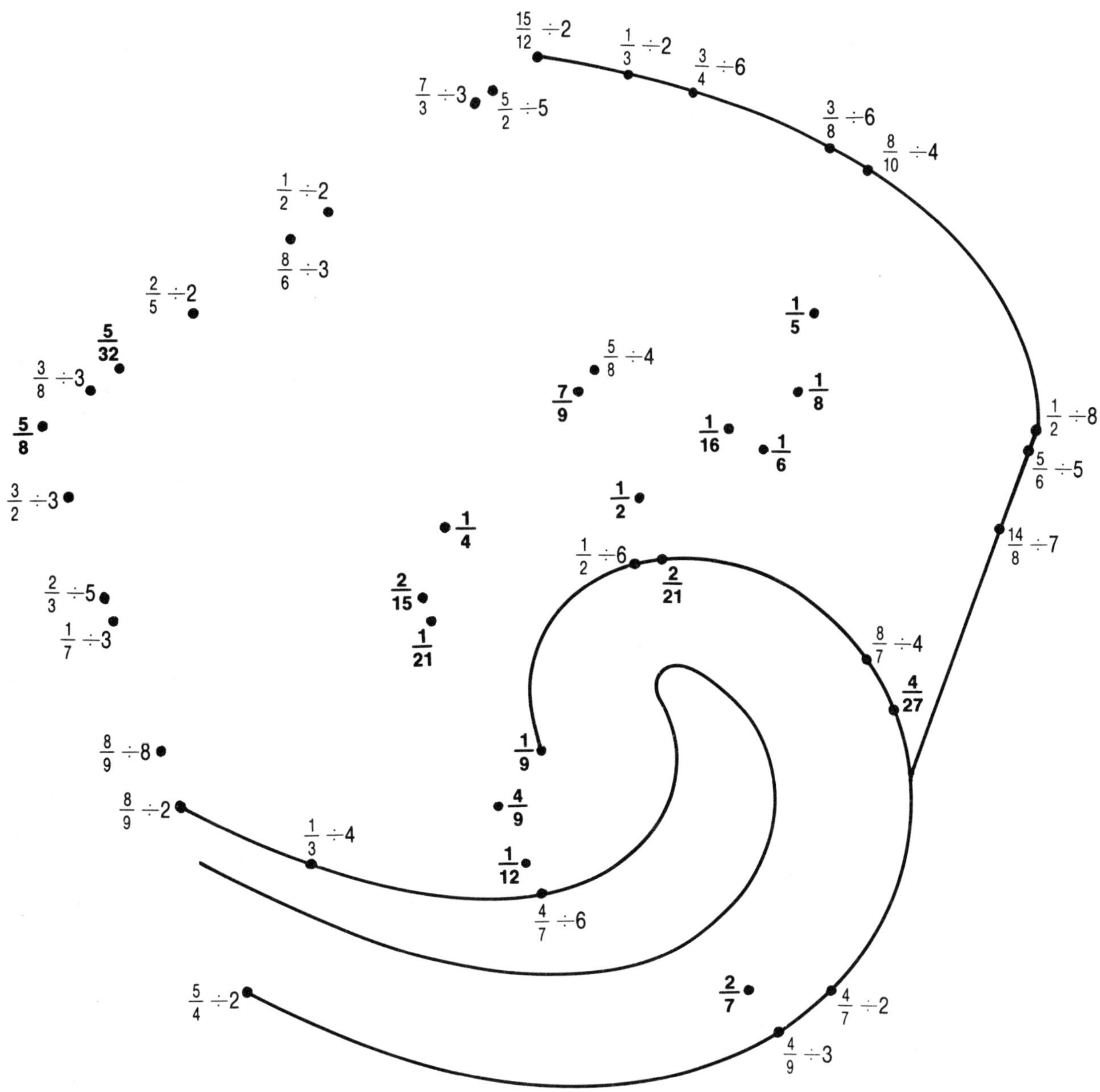

CORRECTION KEY

Designs in Math - Divide Fractions ... Page 17

***Use this design as an answer key.**
***Allow your students to correct their own work.**
Make a transparency of this design and instruct your students to place the transparency over their completed design for a quick and easy check.

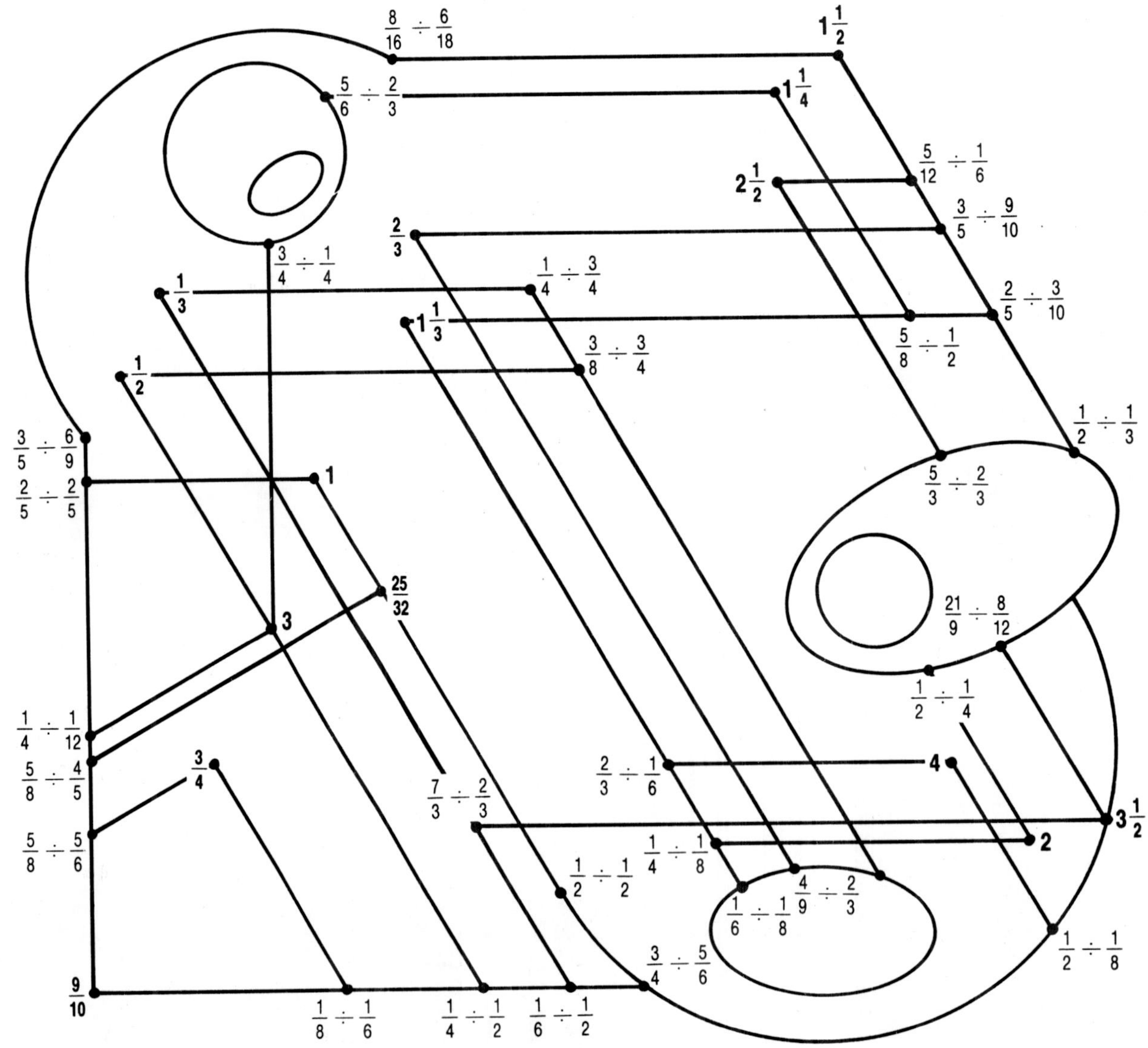

Name ______________________________ **Date** ______________

Read all of the directions before you begin.

1. **Choose one** of the **problems** next to a dot.
2. **Determine** the **quotient** of the two fractions.
3. **Use a straight edge** and draw a line to the **correct answer**. Some answers may be used more than once.
4. **Make sure** you do **all** of the problems. Some lines pass through dots that are **not** part of the one being completed.
5. **You may** want to **color** or **shade your design** after you have completed the problems.

***NOTE:** The answers are **reduced** to their **simplest form**.

EXAMPLE: *Division of fractional numbers.* $\frac{5}{6} \div \frac{1}{3} = \frac{5}{\not{6}_{2}} \times \frac{\not{3}^{1}}{1} = \frac{5}{2} = 2\frac{1}{2}$

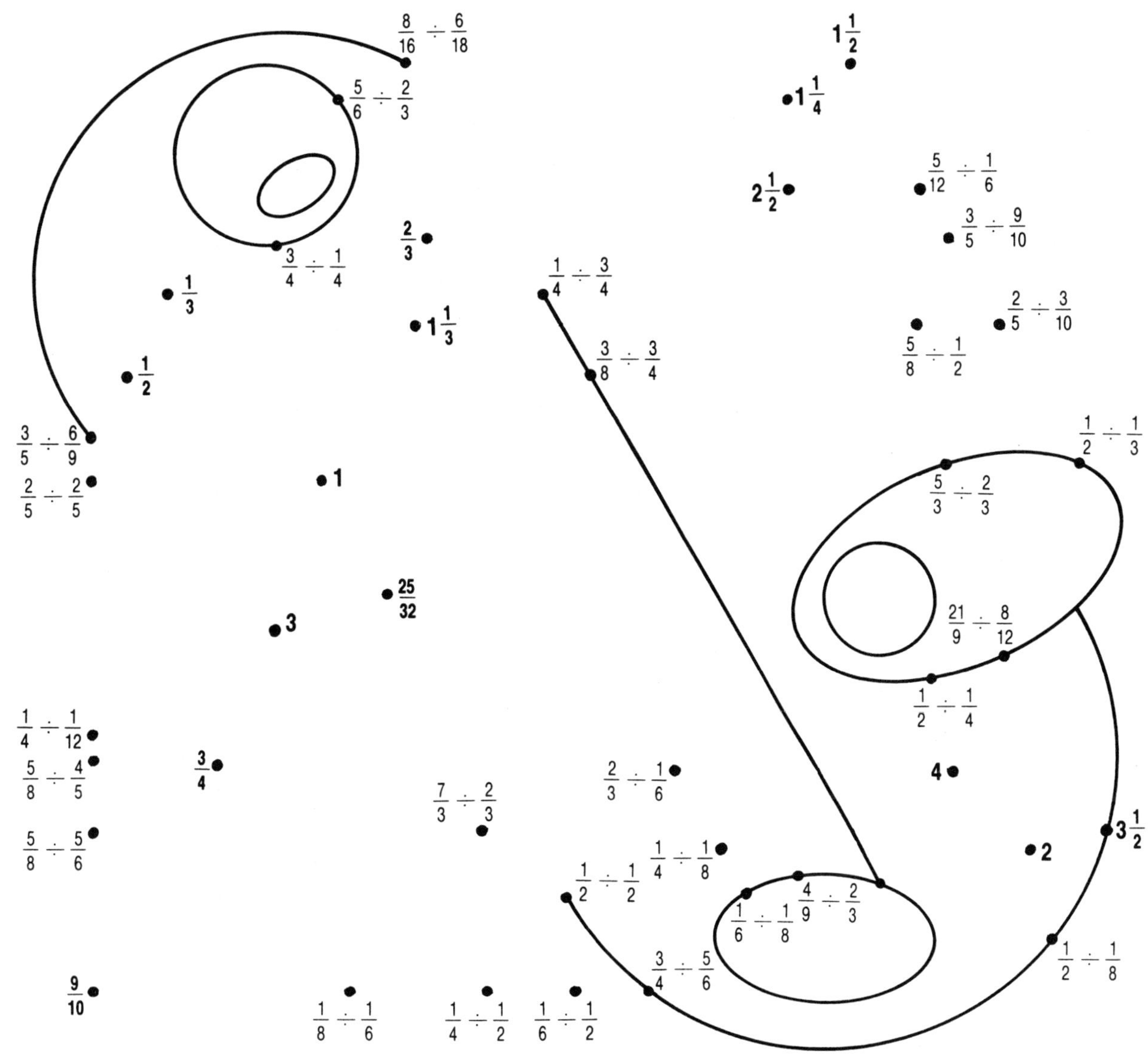

CORRECTION KEY

Designs in Math - Divide Fractions ... Page 18

***Use this design as an answer key.**

***Allow your students to correct their own work.**
Make a transparency of this design and instruct your students to place the transparency over their completed design for a quick and easy check.

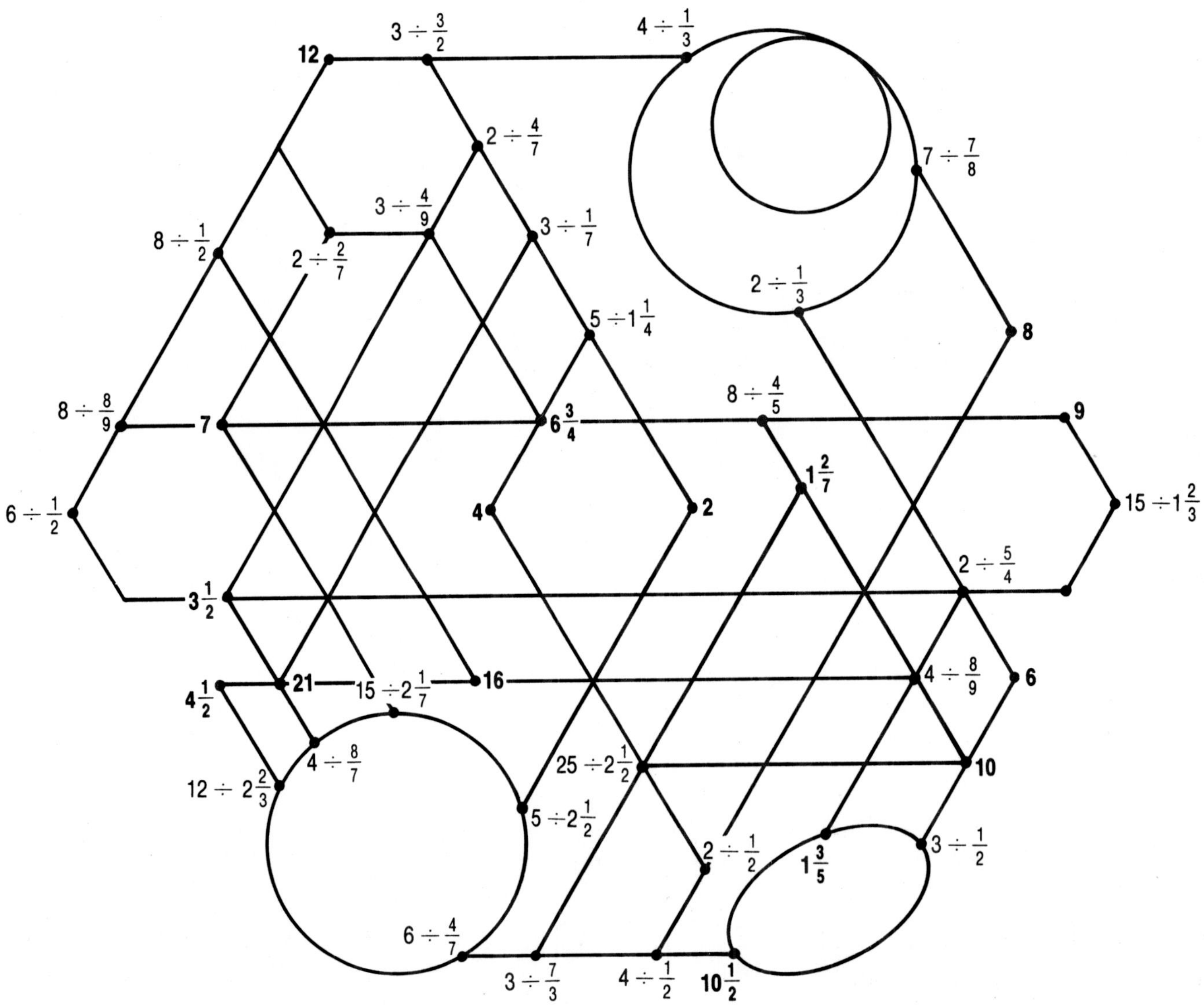

Name ______________________ Date ____________

Read all of the directions before you begin.

1. **Choose one** of the **problems** next to a dot.
2. **Determine** the **quotient** of the fractional number and the whole number.
3. **Use a straight edge** and draw a line to the **correct answer**.
   Some answers may be used more than once.
4. **Make sure** you do **all** of the problems. Some lines pass through dots that are **not** part of the one being completed.
5. **You may** want to **color** or **shade your design** after you have completed the problems.

***NOTE:** The answers are **reduced** to their **simplest form**.

***EXAMPLE:** Division using a whole number and a mixed number.* $6 \div 1\frac{1}{2} = \frac{6}{1} \div \frac{3}{2} = \frac{\cancel{6}^{2}}{1} \times \frac{2}{\cancel{3}_{1}} = 4$

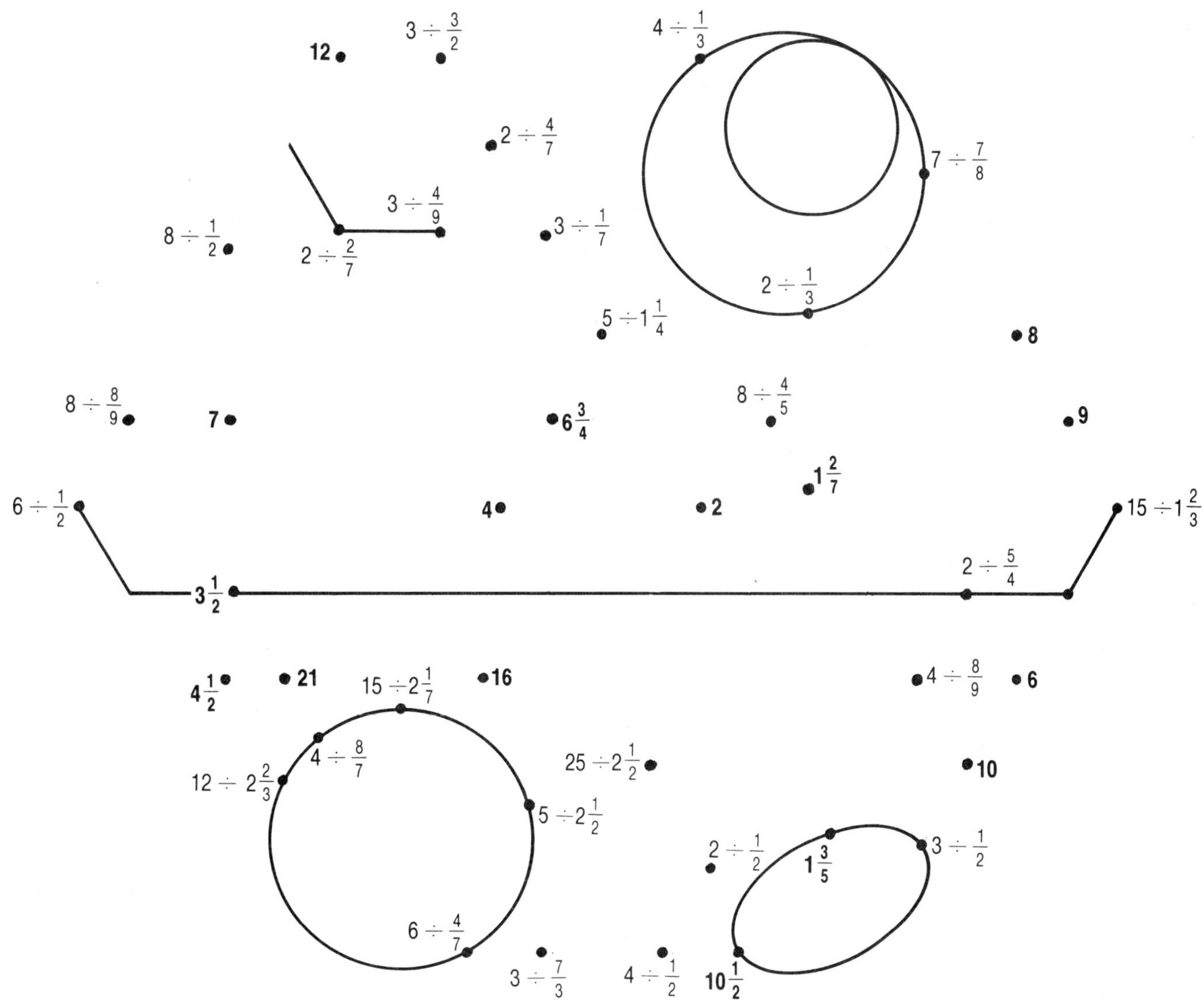

CORRECTION KEY

Designs in Math - Divide Fractions ... Page 19

***Use this design as an answer key.**

***Allow your students to correct their own work.**
Make a transparency of this design and instruct your students to place the transparency over their completed design for a quick and easy check.

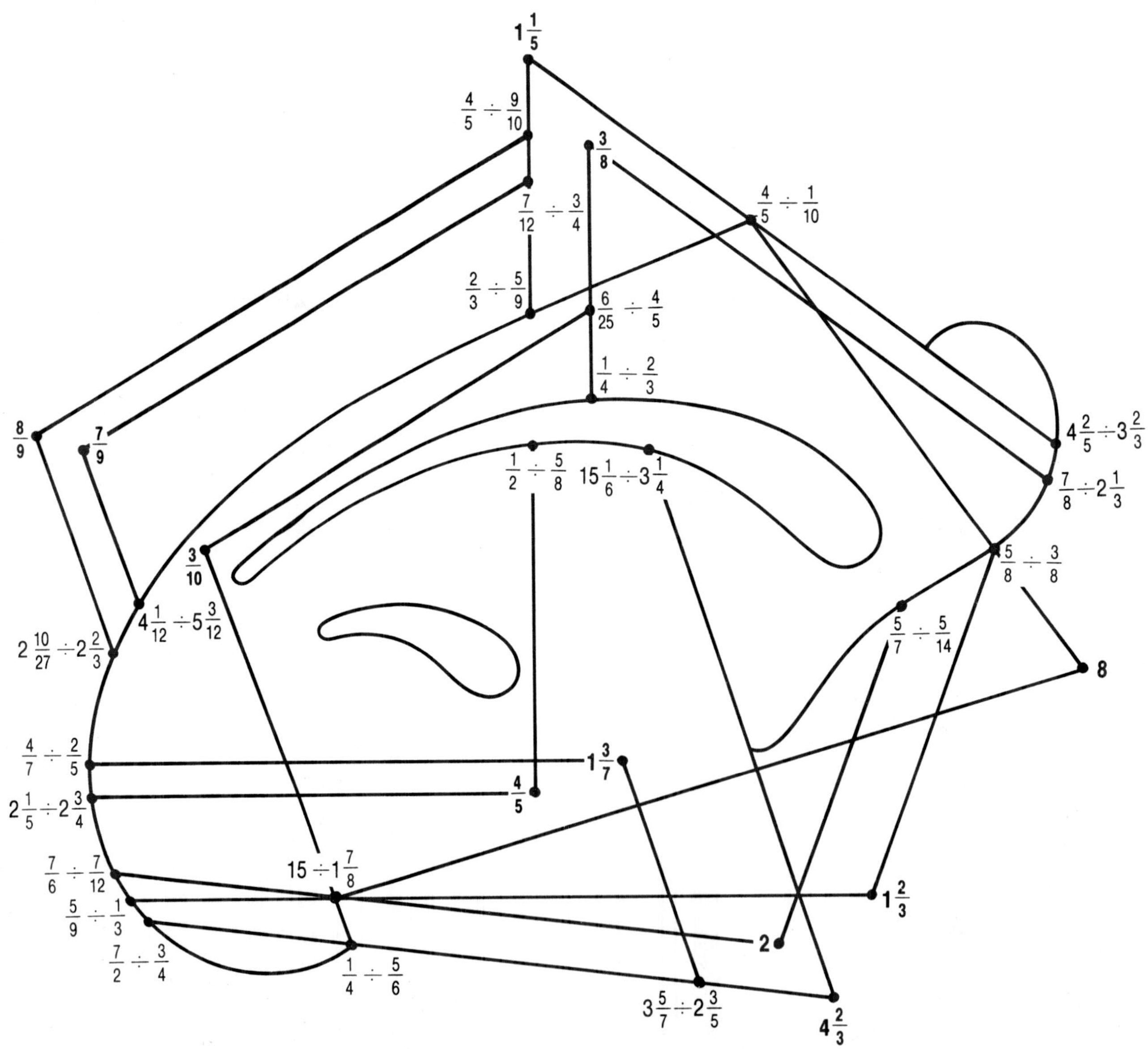

Name ______________________________ Date ______________

Read all of the directions before you begin.

1. **Choose one** of the **problems** next to a dot.
2. **Determine** the **quotient** of the two fractions.
3. **Use a straight edge** and draw a line to the **correct answer**.
   Some answers may be used more than once.
4. **Make sure** you do **all** of the problems. Some lines pass through dots that are **not** part of the one being completed.
5. **You may** want to **color** or **shade your design** after you have completed the problems.

***NOTE:** The answers are **reduced** to their **simplest form**.

EXAMPLE: Division of mixed numbers.

$$2\frac{1}{3} \div 1\frac{1}{3} = \frac{7}{3} \div \frac{4}{3} = \frac{7}{\cancel{3}_1} \times \frac{\cancel{3}^1}{4} = \frac{7}{4} = 1\frac{3}{4}$$

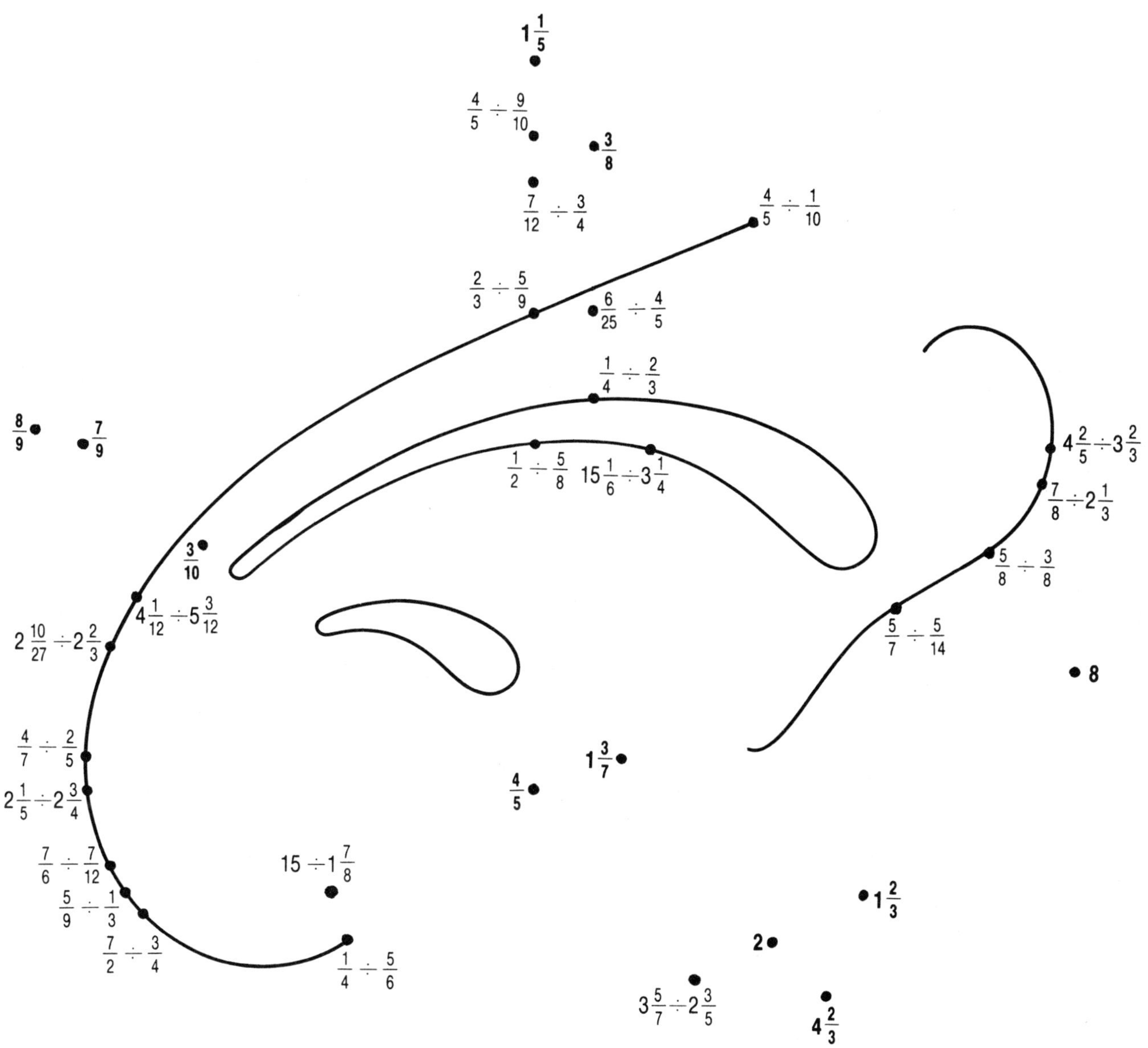

CORRECTION KEY

Designs in Math - Divide Fractions ... Page 20

***Use this design as an answer key.**

***Allow your students to correct their own work.**
Make a transparency of this design and instruct your students to place the transparency over their completed design for a quick and easy check.

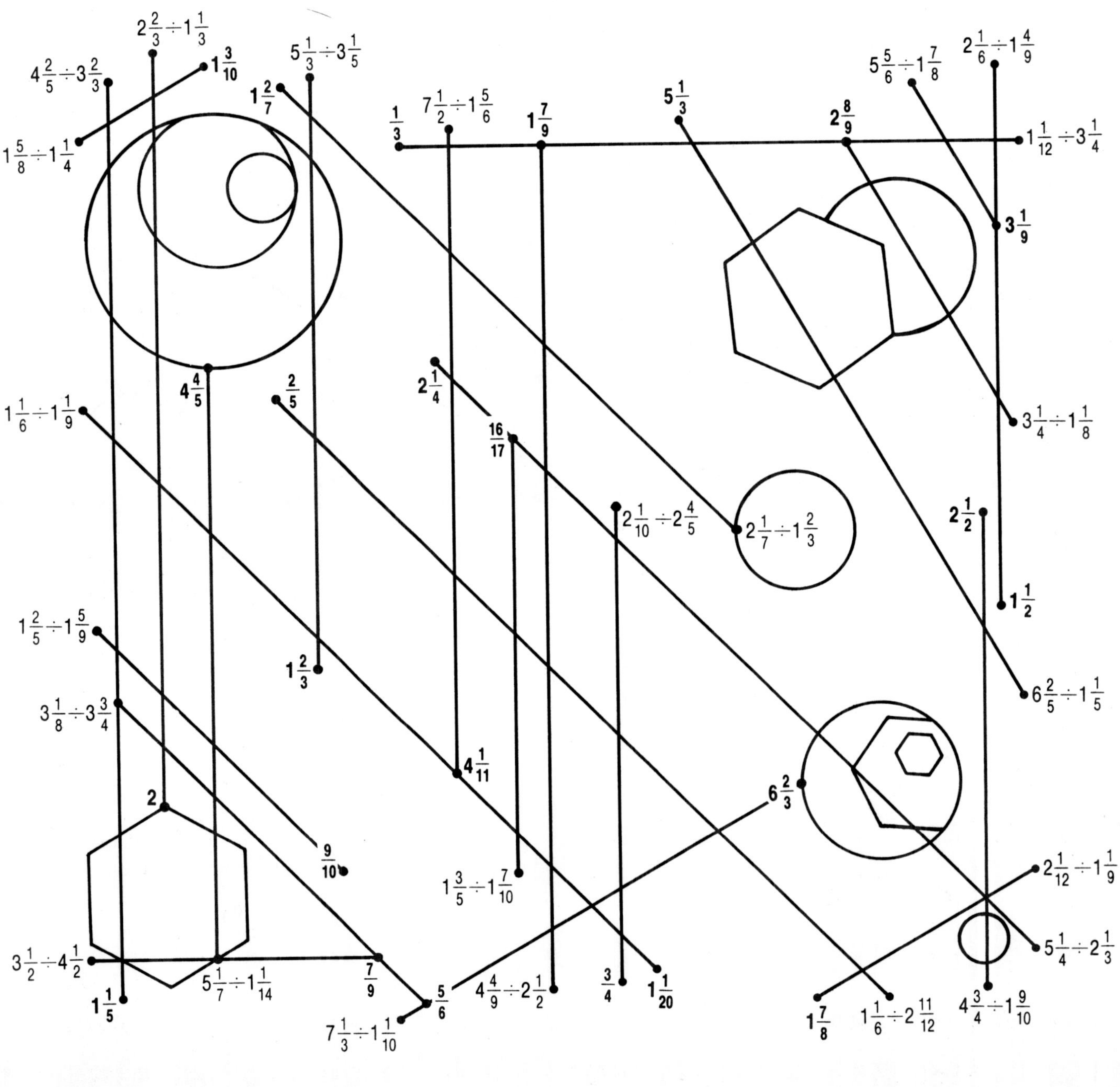

Name ______________________________ Date ______________

Read all of the directions before you begin.

1. **Choose one** of the **problems** next to a dot.
2. **Determine** the **quotient** of the two mixed numbers.
3. **Use a straight edge** and draw a line to the **correct answer**. Some answers may be used more than once.
4. **Make sure** you do **all** of the problems. Some lines pass through dots that are **not** part of the one being completed.
5. **You may** want to **color** or **shade your design** after you have completed the problems.

***NOTE:** The answers are **reduced** to their **simplest form**.

***EXAMPLE:** Division of mixed numbers.*

$$2\frac{1}{3} \div 1\frac{1}{3} = \frac{7}{3} \div \frac{4}{3} = \frac{\overset{1}{\cancel{7}}}{\underset{1}{\cancel{3}}} \times \frac{\overset{1}{\cancel{3}}}{4} = \frac{7}{4} = 1\frac{3}{4}$$

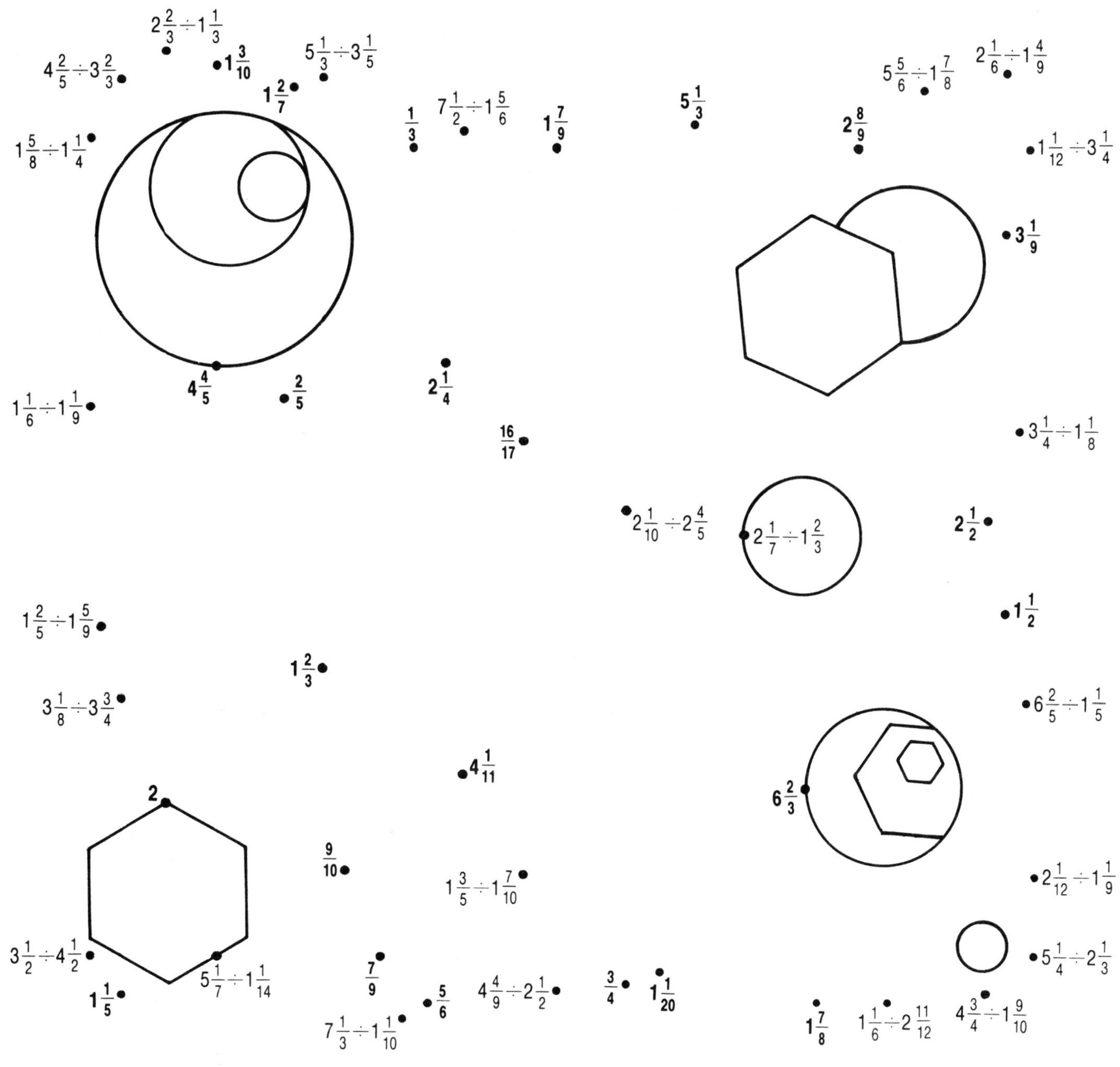

G.E.C. PUBLICATIONS

golden educational CENTER

"LEADING THE WAY IN CREATIVE EDUCATIONAL MATERIALS"™

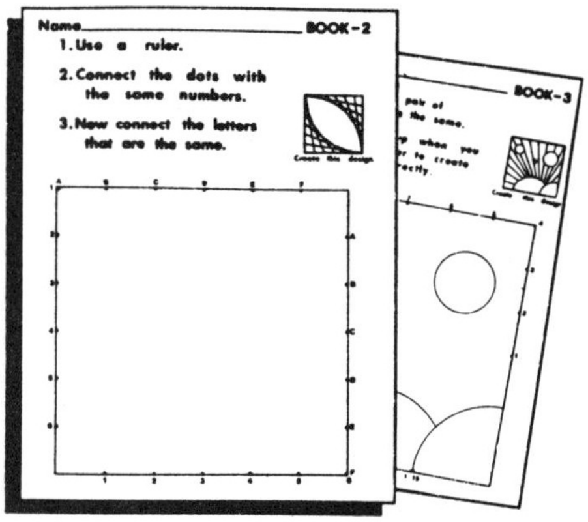

Creating Line Designs

Students use a straight edge to sequentially connect dots in order to create geometric designs. These activities develop visual perception & eye-hand coordina-tion as they build the students' self-confidence. **Your students will love doing these activities!**

GEC-1001	**Book-1**	........K-1st
GEC-1002	**Book-2**	.....1st-5th
GEC-1003	**Book-3**	.....3rd-6th
GEC-1004	**Book-4**	.....4th-7th

Story Problems Made Easy

Paring down to the most essen-tial information, students can now see and understand the most important elements of a story problem. **These might be the greatest stories ever told!**

GEC-1101 **Book 1** - Sums to 20

GEC-1102 **Book 2** - Sums to 99 (No Carrying/Borrowing)

GEC-1103 **Book 3** - Sums to122 (With Carrying/Borrowing)

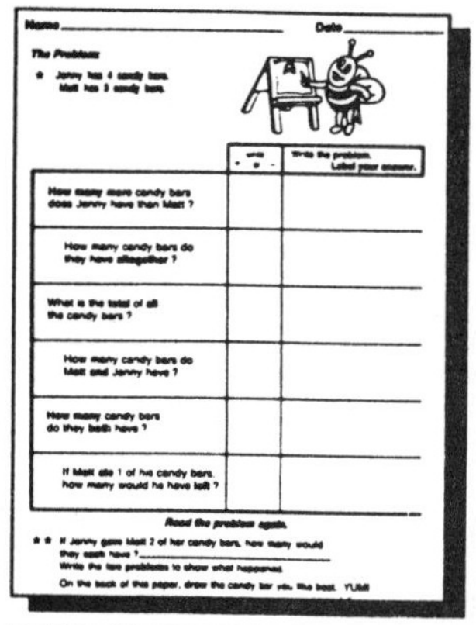

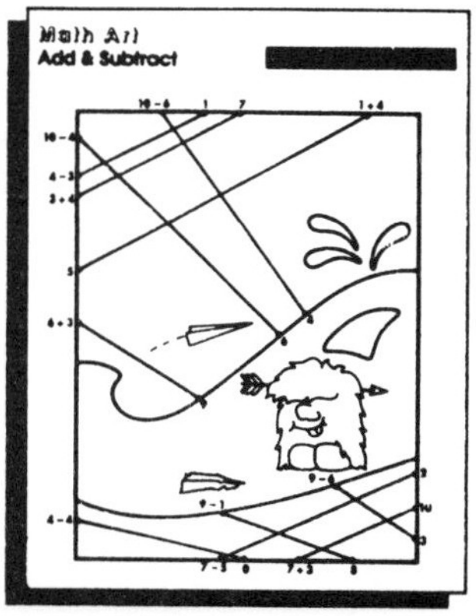

Beginning Math Art

These terrific new books show students the fundamentals of design while they practice solving their math problems. Young students will enjoy doing the 20 activity pages in each of these books

GEC-1013	**Add & Subtract**	0 to 10
GEC-1014	**Add & Subtract**	11 to 20
GEC-1015	**Multiply & Divide**	0 to 12

Designs in Math

Students love to create geo-metric designs by connecting dots and letters between prob-lems and the correct answers. Each book contains 20 pages with one design per page.

GEC-1006	**Addition**.....2nd-6th
GEC-1007	**Subtraction**.....2nd-6th
GEC-1008	**Multiplication**.....3rd-6th
GEC-1009	**Division**.....3rd-7th
GEC-1010	**Fractions**.....4th-8th
GEC-1011	**Equiv. Frac-Dec**...5th-8th

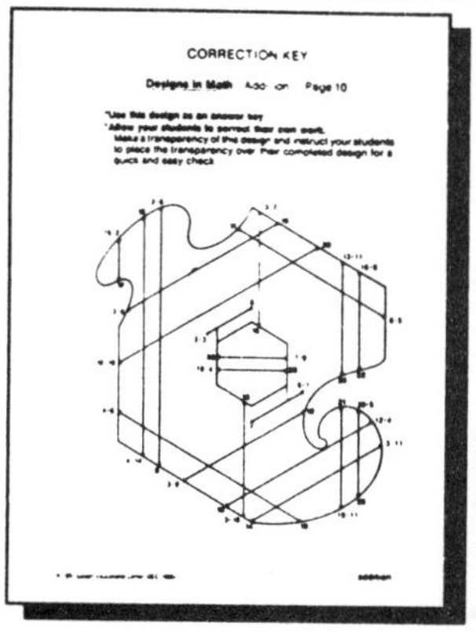

Exciting *New!* Ecology and Art Activities

Ecology Math

Students are presented with simplified ecological information. They then write their ideas on how they can help the ecological situation. The lessons continue with mathematical problems referring to the information first given. Seventeen math lessons with 17 discussion/information sheets in all.

4th and above

GEC-1105

Ecology Activities

Teachers discuss the simplified ecology information given in each of the 18 activities with their students. Children are asked to think of things they might be able to do to help with the problem being discussed. There is an art activity to go with each lesson.

Preschool – 3rd

GEC-1205

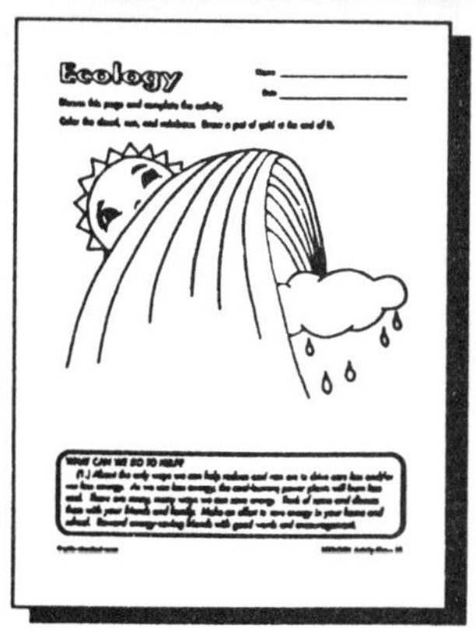

Holiday Greeting Cards

This book has 20 activities where children can create holiday greeting cards for friends and family. Students experiment with several different art media. Teachers love these activities because everything they need is easily obtainable, inexpensive and requires very little prepara-tion.

Preschool – 3rd

GEC-1201

Creating Calendars

Children complete an art activity for each month of the year as well as a calendar grid. They can practice writing their numbers or just cutting and pasting. This book is easy to use and materials are inexpensive and easily obtainable.

K – 3rd

GEC-1202

golden educational center

G.E.C. PUBLICATIONS

"LEADING THE WAY IN CREATIVE EDUCATIONAL MATERIALS"™

Country Study Materials

These books have a section on each of the independent countries of North and South America. Each section has a large country map, a page of current facts and interesting information, a short one or two page history through independence and one page of questions that can be answered with the information we provide. There is also an answer key for each section.

GEC-1965 **North America Country Studies** (96 pages)
GEC-1975 **South America Country Studies** (96 pages)

The layout and information of these books are similar to North and South America. Each country or province has its own respective section. **Your kids will love the easy–to–use format!**

GEC-1935 **Far East Country Studies** (112 pages)
GEC-1936 **Middle East Country Studies** (112 pages)
GEC-1985 **Canada Province Studies** (96 pages)

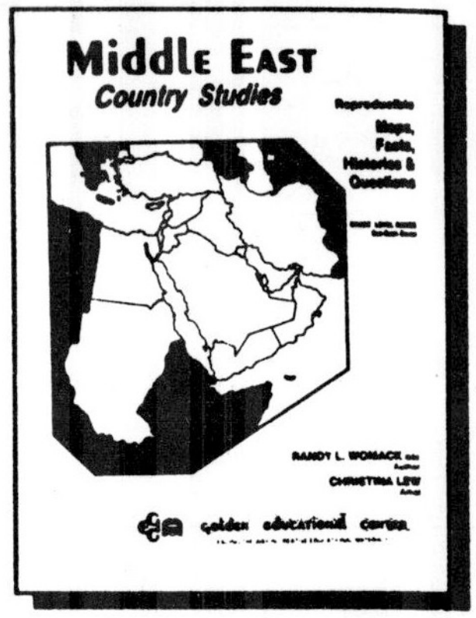

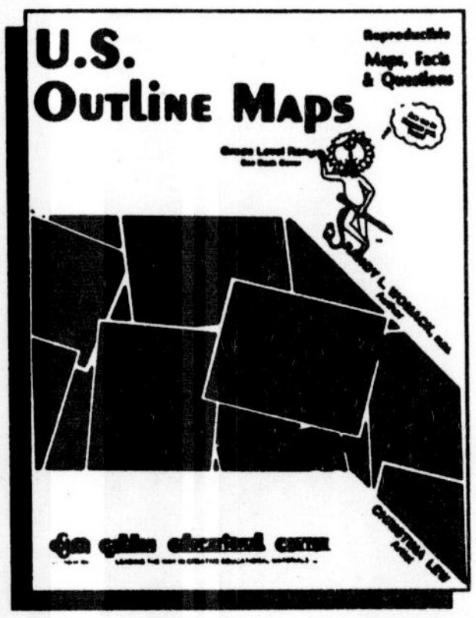

U.S. Outline Maps & State Studies

This 112 page book has an individual Fact Sheet and Outline Map for each of the 50 states, Washing-ton D.C., and the entire U.S. U.S. Waterway and State Bound-ary Maps are also included. The question and research activity pages can be used with each of the states.

GEC-1992

U.S. Geography

The sections of this 80 page book include a World Overview, Physical, Economical, and Political Features as well as Climate information. Maps, activities and questions are included in each section. There is even a review section for the entire book. **This book has a unique, easy-to-use (and read) format that you and your students will enjoy!**

GEC-1993

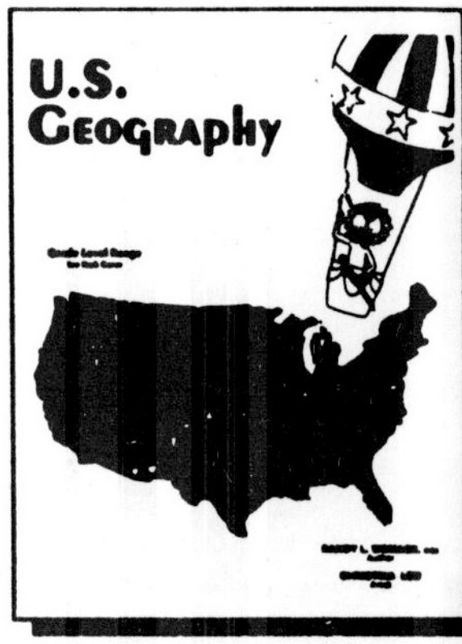

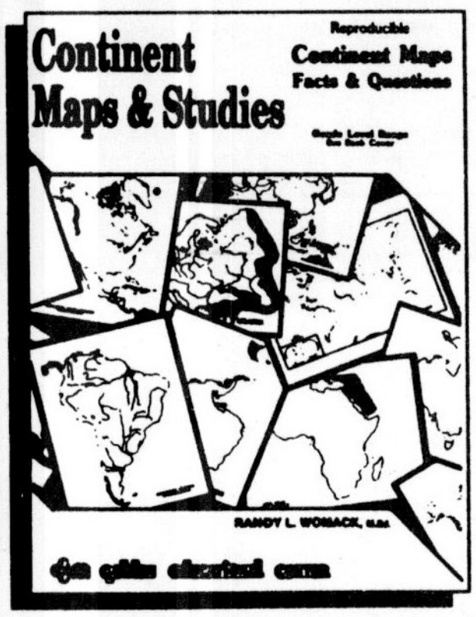

Continent Maps & Studies

This book contains an Outline, Waterway, and Political Boundary Map and Individual Fact Sheet for all of the continents. There are questions, research activities and a glossary that can be used with each of continent section. World Maps and answer Keys are also included.

GEC-1905

Learning the Continents

Africa, Asia, Europe, North America and South America each have a 16 page section of maps & activities. Students use maps to identify, memorize and locate the countries, waterways, and points of interest on each continent. A word search is also included.

GEC-1906

U.S. Word Search and State Studies

This 112 page book has a large word search map for each state. Each state also has an interesting fact and short history sheet that can be used to facilitate more research about the respective state.

GEC-1915